Collaboration, Communications, and Critical Thinking

Collaboration, Communications, and Critical Thinking

A STEM-Inspired Path across the Curriculum

Dennis Adams and Mary Hamm

ROWMAN & LITTLEFIELD
Lanham • Boulder • New York • London

Published by Rowman & Littlefield
An imprint of The Rowman & Littlefield Publishing Group, Inc.
4501 Forbes Boulevard, Suite 200, Lanham, Maryland 20706
www.rowman.com

6 Tinworth Street, London SE11 5AL, United Kingdom

Copyright © 2019 by Dennis Adams and Mary Hamm

All rights reserved. No part of this book may be reproduced in any form or by any electronic or mechanical means, including information storage and retrieval systems, without written permission from the publisher, except by a reviewer who may quote passages in a review.

British Library Cataloguing in Publication Information Available

Library of Congress Cataloging-in-Publication Data

Names: Adams, Dennis, 1947– author. | Hamm, Mary, author.
Title: Collaboration, communications, and critical thinking : a STEM-inspired path across the curriculum / Dennis Adams and Mary Hamm.
Description: Lanham, Maryland : Rowman & Littlefield, 2019. | Includes bibliographical references.
Identifiers: LCCN 2018056377 (print) | LCCN 2019004570 (ebook) |
    ISBN 9781475850000 (Electronic) | ISBN 9781475849981 (cloth) |
    ISBN 9781475849998 (pbk.)
Subjects: LCSH: Education—Curricula. | Critical thinking. | Group work in education. | Communication in education. | Interdisciplinary approach in education. | Science—Study and teaching.
Classification: LCC LB1570 (ebook) | LCC LB1570 .A258 2019 (print) |
    DDC 375—dc23
LC record available at https://lccn.loc.gov/2018056377

Contents

Preface	vii
Introduction: Essential Skills in a Changing World	1
1 Thinking Skills: Critical and Creative Thinking in a Technologically Intensive World	19
2 Collaborative Learning: Teamwork and Social Learning Strategies	43
3 Communications Technologies	67
4 Science and Mathematics: Twenty-First-Century Practices in the Classroom	85
5 Language and Literacy: Communications Skills in a Digital Age	117
6 Arts Education: Connections, Knowledge, and Informed Encounters	145
About the Authors	177

Preface

The STEM subjects (science, technology, engineering, and mathematics) have natural connections to other subjects across the curriculum. To paraphrase the former president of Harvard University Drew Gilpin Faust: A liberal arts education is one that embraces and requires broad learning across the fields of natural and social sciences and the humanities. Also, one of the most important goals of education in a democracy is to ensure that citizens possess useful habits of mind and have the analytic capacities needed to shape the human experience.

Natural realities, technological forces, and new possibilities provide the context for this book's approach to ideas and methods for infusing twenty-first-century skills across the curriculum. When it comes to attention-devouring devices like iPhones, we suggest a cautiously positive approach. An overall basic assumption here is that all students must be educated to the point where they can deal with the problems *and* the possibilities found in a rapidly changing world.

The old foundations of reading, writing, and mathematics, and the new basics of science, technology, and the arts are viewed here as being at the core of schooling. Finding paths between subjects matters. This requires paying attention to how elements of the STEM subjects can be integrated across the curriculum. It also requires that some of the intellectual tools found in other subjects be incorporated into STEM lessons.

The best way to lower the barriers between the arts and the sciences is through continuously informed conversation. Artificial intelligence (AI) is a good example of a STEM-related product that is too important to be left to scientists, engineers, technocrats, or the marketplace. A good liberal arts education is key to working that one out. When it comes to dealing with deep and difficult issues, creative thinking, collaborative learning, technological wisdom, and adaptability all come into play.

In any era, transforming a classroom or school into a contemporary learning environment requires paying attention to current events and challenges; it helps to have consistent classroom routines, using assessment data for improvement, and having a coherent organizational structure. Even younger students can explore the good and bad elements of our tech-enabled lives. Being able to look ahead involves looking in the rearview mirror as well. The best ideas of yesterday and today have a lot to teach us. In other words, it helps to look around today if you want to figure out what tomorrow might look like. Any bridge to the essential skills of the future is bound to be built with expansive thought, collaboration, and the ability to reach for higher levels of understanding.

Broad cultural, social, technological, and educational trends are challenging old assumptions and inviting new dreams. As changes take place, there is a natural tension between what we know and what we can imagine. Recognizing that reality requires exploring topics related to teacher professional development, the cognitive nature of learning, and the pedagogical implications of technology.

In an increasingly globalized world, the next generation has to be more savvy about what is going on worldwide. To make this happen, school culture and lessons have to pay closer attention to an increasingly interconnected world. To be successful in such an environment, schools should be updated in a way that breathes new life into teaching and learning.

The challenges associated with globalization, instructional reform, and new media bring with them the danger of change biting back. Technology is a good example of an area where we have to think critically. Although it's not going to take over the world anytime soon, it is bound change us in unpredictable ways.

It's a complicated and uncertain time with basic technology present in every classroom. The question is, How much *digital* technology *should* be in classrooms? How far down in the elementary grades should student work with digital devices go? To answer these questions requires thinking about the costs, the benefits, and the skills needed before investing in expensive educational tools and materials. At least one thing is clear: digital devices should not be allowed to take time away from the hands-on exploration and collaborative problem-solving that children need.

Change really does favor the prepared mind. Reflection, discussion, and cultivating the disposition for creative and critical thinking can always inform and enrich teaching and learning. It is important to recognize the fact that learning is a continuum; a lifelong awakening to the complexity of the world should be reckoned with.

In times of rapid change, there tends to be a power struggle between what we know and what we can imagine. When schools feel the pressure to change the habit of the familiar, it is tempting to resist and simply continue with day-to-day routines. So it is important to recognize that there are some very negative consequences attached to succumbing to distraction and indifference.

There are no "silver bullet" solutions to educational problems. We certainly can't fire our way to better schools. But we can identify and respect the complexity of the problems that have to be solved. Although schools can't do it alone, they can make a big difference.

In many ways, the whole society is responsible if the home and school environments are dilapidated and in dire straits. For students to be successful, schools and community-based support have to be wrapped around each other in a way that supplies a good educational foundation and sustained support.

The need to move forward to future uncertainties is clear. The days of fairly stationary educational goals are over; hitting moving targets is part of today's reality. We can anticipate constant waves of change flowing across any structure that gets built. One of the few permanent things now is change itself.

Education can be a powerful instrument for social progress in an environment where teachers have the opportunity to forge the future they imagine. With or without outside obstacles, the teacher is the biggest single determinant of how well students move along the road to a deep understanding of whatever subject is being studied. This is the key to quality learning in the classroom: a teacher who clearly understands the subjects they teach and is fully aware of the characteristics of effective instruction.

We can't predict the future, but it is possible to learn about the different terrains where we may all find ourselves living and working. Fortunately, certain guidelines and basic pedagogical principles will not disappear. For example, in any conceivable future, the curriculum has to be meaningful, make connections, emphasize responsibility, and reflect human values.

Collaboration, Communications, and Critical Thinking: A STEM-Inspired Path across the Curriculum suggests some possible routes can be taken today and explores conceptual principles and practical approaches that have meaning for tomorrow's schools. In addition, there are plenty of activities that teachers can use on Monday. Although attention is paid to modern realities, we leave plenty of room for idealism, hope, and practical ideas that will help us all develop schools worthy of our children.

> Nothing great was ever achieved without enthusiasm.
> —Ralph Waldo Emerson

Introduction
Essential Skills in a Changing World

> New ideas lie at the intersection of learning, imagination, tenacity, and reality. What a learner believes is possible helps shape the range of possibilities.
>
> —C. S. Dweck

A person's aptitude for managing the future has a lot to do with his/her skill at understanding the past. Like the best educational ideas of yesterday, tomorrow's curriculum and instruction will build on informed thinking, collaboration, and active learning. Of course, there are times when the debris of the past must be cleared before making changes. But opening some space for creating new visions doesn't mean discarding useful ideas and approaches from the past.

In the last few decades, we have seen major changes in academic subjects, the global economy, and the nature of human interaction. At the same time, educators and policy makers have been creating new standards and methods for encouraging a climate of achievement and creativity in the classroom.

The long-term health of our society and economy depends on creating even more innovation. For future innovators to get it right, they need to go beyond the STEM subjects of science, technology, engineering, and mathematics. Reading, writing, literature, and the arts are examples of other subjects that contribute to building the essential skills needed to succeed in the future.

Developing the imaginative and inventive capacities of young people requires tapping the imaginative possibilities found in subjects across the curriculum. This requires being able to work collaboratively, think critically, and communicate well with peers in the classroom and around the world (Cummins, 2018).

In terms of opening doors to inventive success, much depends on what happens in the space between group members. Some of us may be more imaginative than others, but *everyone* can make creative contributions. The boundary is set by the limits of the human imagination and the natural laws of the universe.

In many ways, the whole is greater than the sum of its parts. Informed teamwork and discipline have more to do with innovation than individual flashes of imagination. After all, lightning bolts don't hit all that often. Genius may speed things up, but being able to work with others to build knowledge and solve challenging problems matters more.

Technology can help or hurt. Love it or hate it, tech products will continue to be important. But they have never been—nor will they ever be—the only thing. Still, tech tools and applications have the potential to be both wonderful and horrifying. They can help us along the path to socialization and deep learning. At the same time, our technological tools serve as a means of government spying, with the use of drones that decide whom to kill, and the privatization of biological life. Obviously, these issues should not be left to technocrats and private corporations.

In today's world, social-educational changes and technological breakthroughs are reaching a critical mass. The question is, How can we design machines, engage students, and pay close attention to the humanities? And, how can all of this come together in a way that serves the rapidly changing world?

SERVING A BRIGHTER FUTURE

In an interconnected world, it is up to educators to conceptualize and construct learning environments that reflect their best hopes, dreams, and values. The next step is for educators and students to symbolically link arms and reach past today's obstacles.

It is impossible to prepare for tomorrow without looking in the rearview mirror. Education has always served as an instrument for progress by providing opportunities to forge the best future that can be imagined. It didn't always work out, but at least it has been shown that it is possible to reach beyond self-interest.

Concern and empathy for others seem a little diminished today. One of the reasons for this is that realities have been altered by ever stronger breezes blowing across both the conscious and the unconscious layers of our minds. An insightful teacher can make a difference by instructing and inspiring thinking and feeling. Considering the possible outcomes of personal and group actions will help both students and teachers deal with the direction of eventful changes.

Change is inevitable and by its very nature, disruptive. Along the path to the future, we have to realize that there are consequences to any human action that cannot be completely understood or fully predicted.

The speed of today's changes has a lot to do with pressures to quicken the pace of educational renewal. Certainly, there are possibilities out there that can help us build a better world. But we have to make sure that human intellect and the products of science serve a brighter future.

Take the example of artificial intelligence (AI). It is already good at performing a single task. But AI has a way to go before it can move beyond simple decision-making to *higher level functions including using predictions to weigh outcomes and pass judgment* (Agrawal et al., 2018).[1]

A note of caution: working alone in front of a screen often doesn't have positive results—especially at the primary school level. Whether it is AI, social networks, or any other computer-based activity, it is important to recognize the advantages of physical proximity. No matter what happens, working face-to-face in community with others will continue to be a powerful motivator.

Although technology *can* sometimes help, active, participatory learning remains the key to understanding content.

New ideas and products often involve taking the tools, concepts, and understandings available today and mixing them together in a way that creates something new. Teamwork skills and information literacy matter because inventive new concepts and imaginative possibilities often depend on local and global networks of information and knowledge (Johnson, 2014).

A message for both student and teacher: get used to teamwork, ambiguity, and lifelong learning. A student's approach to academic work today has many implications for their behavior today and tomorrow. Realize that school is more than preparing for life; it *is* life.

Academic success depends more than ever on intellectual curiosity, human nature, and student/teacher ingenuity. Across the curriculum, the keys to academic success are teamwork, asking good questions, and generating new ideas. Also, information and digital literacy are central to helping students understand everything—from the arts to current events.

CREATIVE THINKING IN AN UNCERTAIN FUTURE

The effectiveness of new educational approaches has a lot to do with the strength of the people behind change and the strength of the school culture (Gruenert and Whitaker, 2015). Consider arts education where new conceptual challenges and techniques abound. Like it or hate it, the Web does allow you to tour the best museums on the planet and even look inside an artist's studio.

Advances in digital technology and the Internet have created new kinds of art possibilities that can help make good art or ruin the entire process. Man Ray, Alexander Calder, and Igor Stravinsky all used the new technologies of their time to conceptualize new forms of visual art and music. The difference is that with today's technology just about anyone with a computer can hammer out an art print or a musical composition. How good are you at sorting out the good from the horrible?

Take the example of music. Today's technology is so powerful that students who don't know anything about scales—or much else musically—can use devices and apps like the *Native Instruments Maschine* to make original compositions. For children, it may be fun and informative. Still, melodies, timing, lyrics, and just about everything else can be worked out by the computer as students bang away on the keyboard.

Teachers and students have to figure out when it's best to use basic technology (like paint and canvas) and when it is best to use digital tools as assistants in the creative process. There is no reason why anyone has to rush headlong into the new digital age. But no matter how you approach it, neither teaching nor learning should be a passive or a solitary experience in front of a screen.

Neither teachers nor the arts are in danger of being replaced by new forms of technology—or anything else. Digital technology is, at best, a complement (not a replacement) for traditional art, teacher innovation, and student engagement. Science simulations, virtual museum visits, and word processing are one thing; but if tech tools aren't well suited to the task, then they may not be worth the time or the cost.

There is at least some truth to the idea that you should be careful about how you interpret the world because that interpretation may alter the shape of reality. Now artificial intelligence, advanced robotics, and a host of other innovations are replacing jobs higher and higher up the skills ladder (Brynjolfsson and McAfee, 2014).

We have reached a stage where innovation and increases in productivity increase the wealth of fewer and fewer people. As far as the schools are concerned, the wage gap and housing gap have a lot to do with the education gap. Clearly, winner-take-all capitalism leaves many teachers and parents in the dust. The good news is that even first graders realize that the little red hen should at least think about sharing a few of her hard-earned assets.

KNOWLEDGE IN THE SERVICE OF CREATIVITY

Creative thinking involves intellectual curiosity, an openness to new experiences, a vivid imagination, and a spirit of inventiveness. Of course, a certain level of self-discipline always helps. Innovative ideas or products are rarely dreamed up by someone working in isolation. Teamwork and the ability of all concerned to respond to a rapidly changing world have a lot to do with individual and group success.

When it comes to individual teachers in the classroom, turning imaginative possibilities into helpful realities depends on preparation, peer collaboration, educational vision, and a well-designed curriculum. For both teacher and student, risk-taking and the ability to overcome mistakes are part of the process.

Knowledge in the service of creativity may open up some positive, long-term possibilities. For example, as part of hands-on lessons—from the arts to engineering—students can be encouraged to take virtual tours of museums, musical or theatrical productions, and science simulations. (Suggestion: in the classroom, a few online hints are fine for developing questions, but viewing longer segments are best done as homework.)

Networking possibilities are expanding rapidly; today's technology supports global observation, interaction, and learning as never before. It's a new world order. Still, the shape of a nation's future depends more on the quality of its educators than on new media possibilities. Quality instruction comes down to teachers and how well they understand the characteristics of effective instruction (pedagogy). Of course, knowing something about the subject being taught is part of the equation.

Still, stimulating and aesthetic learning environments can sometimes be informed by combining new information distribution possibilities with the more personal, collaborative aspects of interactive face-to-face learning.

Sometimes, mastery of the natural world goes hand in hand with significant percentages of the population trying to avoid things that are difficult to understand or control. Climate change is just one example. Dealing with new ideas and the unknown is not always popular. Although subjectivity may be less common in topics connected to math and science, it can be found in all human endeavors.

In a world of uncertain challenges, good preparation includes developing an analytic spirit, learning to deal with ambiguity, and cultivating the capacity for questioning and judgment. The capacity of teachers and students to deal with change, learn from it, and help each other manage the surrounding uncertainty is critical to the future development of our society.

Change may be influenced by the intellectual energy of a civilization, but you can be sure that it will favor the prepared mind. Learning something new, like much of life, still has its mysteries. Preparing students who can be successful in an unknowable future requires teachers who can help students collaboratively perceive, analyze, interpret, and discover a whole new range of meanings (Beghetto et al., 2014).

SOCIAL REALITIES IN THE AGE OF INFORMATION

When all students have access to computers and other digital devices, the learning gap between students may actually increase. Whether it is on or off campus, after some initial enthusiasm, students who are having academic difficulty tend to waste more time on digital devices than students who are doing well (Pinker, 2014). It seems that providing Internet-connected computers to students who are doing poorly can cause them to go from bad to worse.

Many of our digital tools, like the Internet, have only become popular in the last thirty years or so. Our relationship is still evolving, so it is hard to figure out exactly what's going to happen just over the horizon.

Is digital technology a useful tool, a distraction, or both? A lot depends on how tech tools are used and how students are taught to understand and create with the most powerful media available. If children spend nearly half their waking hours in front of a screen (without adult supervision), bad things can happen. Still, when technology is used well, it can support creativity, collaboration, and curriculum standards in ways that reinforce all three.

Information economies require higher levels of and more frequent education for everyone. There is a convergence in knowledge-producing possibilities involving publishers, schools, the Internet, television, libraries, universities, and museums.

We face a future where human communication, interaction, and learning are no longer bound by time, space, and form. Fortunately, everything we know and value won't fade into oblivion. In any conceivable future, children will have to learn how to read, write, and do mathematics. Also, learners will have to develop the thinking skills needed to sort through today's avalanche of information and take an informed position on a wide range of issues.

For future educational models to be effective, they must be influenced by the research on instruction, new understandings about the social nature of learning, and advancing technological possibilities. Like it or not, educators will bear some of the responsibility for arranging new instructional models in a way that helps produce citizens who can live and work productively in an increasingly complex world. Later chapters will provide more information on this topic.

Every forward-looking educational plan for the future requires putting more resources into focused learning opportunities for teachers. Why? To begin with, it takes well-educated teachers to help students to intelligently connect to wider and wider circles of knowledge and social issues. So, it is little wonder that all current models of curriculum, instruction, and educational development include perpetual staff development and school-embedded learning.

The world outside the classroom is changing more rapidly than ever. Although the pace of educational change is slower, the same can be said for the high-quality schools. Everyday reality intrudes into schools as never before. In today's social milieu, schools have a major role to play in providing socially and intellectually stimulating work for students.

Giving children an education sufficient for healthy and productive lives is more important than ever. The recipe for making this happen doesn't always involve full-scale transformational change. There are times when gradual but constant change is the best way forward. This is especially true when the complexity of social-educational systems makes it difficult to predict or understand the consequences of certain policies and actions.

Question: Are the uncertainties we find around us simply a reflection of our imperfect knowledge of human nature and the natural world? Some things are quite clear. For example, we know that the social nature of language and creative development are so interconnected that it is hard to separate one from the other. Across the curriculum, when new things materialize, their usefulness depends on whether or not they empower us to collaboratively reach further in the world.

To deal with real-world problems and stay relevant, teachers need to know about what's going on in various subject fields. And, they need to work together (i.e., professional development) to translate the implications of current events into terms their students can understand. It may take more than what teachers can do alone to generate adequate resources for integrating the most powerful approaches and methods into day-to-day lessons. Both the pedagogical and funding pieces have to be in place if major changes are going to go smoothly.

To paraphrase Goethe,

What you know or dream you can begin.
Boldness has genius, power, and magic in it.
Engage, then the mind grows heated —
Begin it, and the work can be completed!

SCIENTIFIC AND TECHNOLOGICAL ACCELERATION

Expressions of scientific knowledge and better technology are now global in their effect. Science and its technological tools have directly accounted for fundamental changes in the world economy and, at the same time, tied the world together by nearly instantaneous communication. Still, there is much enthusiasm, controversy, and criticism.

We try to reclaim the middle ground of this love-hate relationship with technology. To paraphrase David Auerbach (2018), many are debating the perils of social media, and some vilify technology as an instigator of our social ills, rather than a symptom. He goes on to suggest that there is something about our digital life that seems to inspire extremes. It often starts with early enthusiasm and utopian fervor; as time goes on, things often collapse into fear and recrimination.

As Auerbach points out, our relationship with technology is still evolving, yet throughout these cycles, we are increasingly intimate and ever more entwined and interdependent. In this book, we try to be balanced and judicious in our approach to the subject.

From genetic engineering to the Internet, technologies and their scientific associates are now in a position to direct and manipulate the world more than ever. As far as the future is concerned, the best we can do is get a dim glimpse of the world ahead by looking around us now.

Genetics, robotics, and nanotechnologies are just three of the newer technologies that pose a potential risk to the physical world. When it comes to technology, there is a mix of positives and negatives. For example, new gene editing techniques can cut out disease-causing sections or build designer humans. Welcome to the twenty-first century, Mary Shelley (the author of *Frankenstein*).

Machines that can self-replicate themselves is another challenge. From the genetic shaping of individuals to digital experiences bought off the shelf, it is becoming easier to create natural facts (like new species or molecules). What are the consequences of giving more and more of our lives over to computer-based algorithms?

Developing a moral compass is more important now than ever. John Goodlad, for example, has for years pointed out that schools are moral enterprises—they are central to the future of our democracy (Goodlad, 2016). Social development and good citizenship are part of today's academic menu. New media can help and/or hurt the process.

The Internet and its "clouds" (like Apple's iCloud) are examples of how, right or wrong, the collective unconscious is a powerful part of moral development. The *cloud,* by the way, may be thought of as distant computers that provide software, data storage, and other information technology services that can be accessed from any Internet connection. The dynamic and compelling forces of information and communication technologies are becoming more effective at creating virtual realities. These made-to-order worlds are bound to take time away from face-to-face human interaction and ordinary sensory experiences. So, be sure to leave space for human interaction.

The optimistic and self-congratulatory high-technology industry may have to step back from its chaotic advance into an uncertain technological future and debate the issues surrounding the inherently negative possibilities of rapidly evolving technologies (Catmull, 2014).

Schools can temper some of the harsh edges of digital technology by becoming good role models for connecting the social aspects of learning with broader online communities. Remember, all instruction works best when it is student centered and designed to facilitate instruction through interactive mediation processes.

The revolution in the provision of information is a huge educational and cultural change that requires new ways of thinking and learning. To use new media and their applications is to extend the idea of what is possible—altering the way people think and the way they act. For a jump of comparable importance, you have to go back to the transformation of culture, thinking, and learning that was caused by the introduction of the printing press.

THE INTERSECTION OF TRENDS, EVENTS, AND LEARNING

If you are reading this book, you are most likely committed to designing engaging learning experiences that can meet the needs of all students. The degree to which you choose to make use of new media is a professional and situational choice. But no matter how much use you make of digital technology, it *is* important to be aware of what it can and can't do.

Ask yourself at least two questions: *How might digital devices and their applications help accomplish instructional goals? What are the existing possibilities and dangers found in the STEM subjects?* Whatever your answers, there is no escape from the mysterious nature of our innovation-driven world.

Whether it is science, technology, engineering, or technology, new perspectives can be opened, illusions magnified, and obstacles to our aspirations revealed. Any of the STEM fields can be dangerous if left in the shadows (Wagner, 2012). A good example is the Internet, where thoughtful quality is hard to find and lunatics are numerous. Books and newspapers (with well-informed writers and editors) are one thing, but being able to sort through an unfiltered glut of online information is quite another.

Genuine face-to-face interaction with others is needed in learning any new idea, approach, or product. That's because genuine learning depends on personal interaction among students and ideas in an aesthetically and intellectually stimulating environment.

Technology, economics, education, and culture are increasingly tied together. Major social and technological advances have moved rapidly—tying humanity together with a new immediacy and intimacy. More than ever, global change is driven by new ideas and media innovations that have taken on a speed of their own. What might get in the way of schools adjusting to these changes?

Test taking and test preparation are but two examples of how efforts to improve the schools can sometimes have the opposite effect. Increasingly, teachers and students are being coerced into spending a lot of time and effort on out-of-context tests. Would it be better to spend the time with more stimulating information and more imaginative activities?

The word *inertia* is often associated with attempts to change the curriculum. Teacher attitudes, assessment practices, the need to reeducate students, and traditional ways of doing things, can all get in the way of initiating a new curriculum while operating an old system. Still, some schools have worked their way around these very obstacles.

In the classroom, students learn best when there is opportunity for social interaction that encourages them to collaboratively create and communicate meaning. This can be done with an array of old and new media that provide information, stories, and folklore in a way that contributes to a shared intellectual curiosity. Therefore, one of the most important features of a successful educational system is providing the capacity for self-renewal and continuous change.

Although teaching and learning have taken on new dimensions, some schools have not changed as much as the world around them. Unlike the tech industry, for example, innovative thinking and novel ideas have often been viewed as risky in the educational field. This is in spite of the fact that purposeful risk can actually increase teacher innovation, get students more collectively engaged, and help all concerned do a better job of dealing with the unknown.

The degree to which you choose to make use of new media is a professional and situational choice. But no matter how much use you make of digital technology, it *is* important to be aware of what it can and can't do.

INNOVATION AND LEARNING IN ASSOCIATION WITH OTHERS

Both innovation and learning are more powerful when done in association with others (Dance and Kaplan, 2018). From Edison to Jobs and others, bringing together creative teams was, in some ways, a greater achievement than the products of any one individual's imagination.

Positive educational change has usually been centered around collaborative teachers and individual schools. Experts and professors can help. But a note of caution: passing legislative bills and drafting school district policies are not going to make all that much difference if teachers don't buy into them.

Whether it's high-tech or low-tech, it is important that everyone in the school community should at least be aware of both intellectual and technical tools in their most powerful forms. It is important to recognize the fact that many subjects are getting so complex that their consequences are getting harder and harder to foresee.

When it comes to quality teaching, it all depends on teachers being up-to-date with the subject matter *and* being familiar with the art and science of teaching (pedagogy). So it should come as no surprise to find out that integrating curricular reform with professional development is one of the keys to successful educational change.

There are solid examples of successful practices and positive educational change around the country. Unfortunately, things that work well in one school may not work so well in another. So, it is up to educators to sort through, select, and adapt the best practices for different school situations and contexts. Administrators and policy makers need to provide the resources that allow new ideas to blossom.

Mistakes are fine, but you will make fewer of them if you learn from the mistakes of others. If you're going to fail on your own, it's best to do it quickly and get over it. And remember, occasional failures are part of the cost of trying to do something unique; just try to learn from them.

To paraphrase Thomas Edison: *Results! Why, man, I have gotten a lot of results! I know several thousand things that won't work.* Like Edison, if you can't find a way to get through a roadblock, it may be possible to make a new pathway to get around it. Just keep at it; tenacity has always been one of the keys to innovation.

Margalit Fox got tired of her job after writing more than 1,400 obituaries for the *New York Times* (Fox, June 28, 2018). In addition, she feared that her own epitaph would say little more than *she changed 50,000 commas into semicolons*. As she left the Times copy desk to write and edit her own work, Fox pointed out that we all have the potential to put a few wrinkles in the social fabric. So give it your best shot.

BECOMING A MORE EFFECTIVE TEACHER

Experienced teachers understand that teaching is a complex undertaking that requires time. Conditions may be difficult, but effective teachers accept and enjoy the challenges. They know that they make a difference in students' lives.

Many teachers describe success in personal terms and view it as related to being personally durable and capable. Knowing the child and understanding the community are frequently mentioned as important. Most teachers who have elected to remain in their schools for more than three years feel in control of their environment and believe teaching is a rich and rewarding experience.

Effective teachers are more likely to believe the following:

- Creating a feeling of excitement about the subject matter or skill being taught is important.
- Children can always learn more and the teachers' effort and energy are instrumental in students' learning.
- Providing children the opportunity for active participatory experiences is a powerful incentive for learning.
- It is important to reflect a strong sense of personal caring about students and adjust instruction to their needs.
- Children try hardest when they are fairly certain of success, but not absolutely positive.
- Students learn most from teachers who believe that the level of student effort can predict achievement.
- Young people learn most when their questions and learning activities are connected with big ideas, key concepts, and their intellectual curiosity.

CONFIRMING THE IDEALS OF PUBLIC EDUCATION

Global competition is one thing, the future of our democracy is quite another. They do, however, have at least one thing in common; both depend on the revitalization of the public schools (Darling-Hammond, 2010).

No matter what happens just over the horizon, the majority of American children will continue to be educated in public schools—making choices about life and connecting to youngsters from different backgrounds. Public school is also where images are formed of what it means to be a good person, having a good life, and living in a good society.

The interpreting and reinterpreting of the ideals of universal public education will be with us throughout the twenty-first century. Small-scale, private-sector experiments are fine, but to reach the majority of students, innovations must be transferred successfully to the public sector.

In the workplace or in school, specific skill and general intellectual development activities completed in small groups can make learning more lively and interesting for everyone. The best schools are constantly building on the latest research-based techniques—while continuously recycling the most successful practices of the past.

Fortunately, many of the basic pedagogical principles remain fairly constant. Whether it's yesterday, today, or tomorrow, effective teachers have been and will continue to teach in ways that help students deeply understand content.

In any conceivable future, educators will be designing activities that emphasize reasoning, collaboration, and communication (Opitz and Ford, 2014). Another consistent educational goal is the creation of learning communities that encourage students to become active and collaborative participants in the construction of meaning.

Some schools really have their act together. But there is general agreement that in at least a few schools, students are made to feel unwelcome, intellectually inadequate, uncomfortable, and bored. At the earliest opportunity, they drop out. For some who stay, schools may offer little encouragement for those who have talents extending beyond the ability to manipulate words and numbers.

Figuring out what to keep the same and what to change isn't easy. Yes, transforming our schools will occasionally require dealing with state and district systems that are hostile to change. But top-down mandates usually do more harm than good.

Large-scale educational change requires sustained public commitment. Once this is in place, we must involve a widespread cadre of public school educators who are willing to implement high-quality approaches in their schools. The overall goal should be nothing less than making sure that all of our schools shine when compared to other schools around the world.

Ideally, curriculum designs should build on Western traditions, while valuing values and cultures from around the world. This has to happen while schools carry out rigorous self-examinations, incorporate vigorous innovations, and develop a commitment to greater effectiveness.

EDUCATION, CULTURE, AND LEADERSHIP

Culture can be viewed as a coherent system of attitudes, values, and institutions that influence both individual and group behaviors. The idea of culture has become ever more elastic and blurred by modern communications, swift transport, and the breakdown of some traditional societies. In some regions of the world, the fabric of civil society is unraveling—making anarchy more common than a coherent ideology. In other areas, the national identity and cohesion are strong.

No matter how enlightened the civic structure is, in a connected world, learning goes beyond what's happening in school. Technology often creates alternative realities and new anxiety multipliers. Entertainment, spectacle, and politics have merged in ways that cause people to have trouble distinguishing lies from the truth. Worse yet, many don't seem to care much about the distinction.

Facts and logic matter. The same can be said for the ability to figure out the difference between what's real and what isn't. So it is little wonder that some educators are concerned about whether or not their students are learning how to draw a clear line between fact and fiction.

Technology can help or hurt. On the positive side, it can present possibilities for weaving the world together as never before. Like books or any other media, whiz-bang technology can be used in smart ways to support learning. On the other hand, the same technology makes it possible to find fearful, dark corners or time-burning distractions.

Cultures that nurture the human creative capacity across age groups usually do better than those that don't. When education is thought of as a continuum from prenatal care through adult life, it is bound to have a more powerful effect. Learning is now much more than something that happens to the young; today, lifelong learning is a fact of life for all of us.

At any level, peer culture and the media can get in the way; but parents are a key ingredient in making sure their children's education is going in the right direction. Academic achievement is strongly influenced by the level of insistence on the part of parents that children take their studies seriously. In today's world, some parents may not be able to help much. They need social support networks and assistance in learning about how child-rearing affects schoolwork.

Fundamental changes in an educational system work best when there are corresponding changes in communities and the larger society. The whole process requires asking hard questions and dealing with some uncomfortable issues. Parallel changes in cultural beliefs, social incentives, and the status of teachers are needed if schools are to make a major impact on childhood learning difficulties.

Limiting the educational focus to concerns like national testing and school choice avoids the more crucial issues of moral numbness, individualism degenerating into greed, spiritual alienation, social injustice, and the diminishing prospects for a healthy future. Zip codes (housing locations) should not be the determining factor of educational success. For too long, many have not worried enough about future generations or obligations that different age groups have to each other.

When it comes to improving schools, it isn't just about money; it also involves values, willingness to sacrifice, and the ability to look at children and young adults in their totality. Also, to deal with many of the issues that are most important to a child's future, educators will have be ready to teach in ways that haven't been fully conceived of yet.

School culture must encourage teachers to collaboratively approach new ways of thinking and turn possibilities into reality. It may not be possible to change yesterday's mistakes, but it is possible to make a positive influence on the future. High-quality teachers and informed parents are the key. But it also takes leadership that is willing to address the issues of our time in a bold manner—to inspire, instill, trust, and carry out carefully conceived plans.

As far as school principals are concerned, success is often attributed to having more latitude in curricular and spending decisions. Other positive factors include having tangible goals, careful teacher recruitment, and parental outreach programs.

DO YOUR BEST TODAY; DO BETTER TOMORROW

Schools can become a community of learners by creating a caring atmosphere, attending to student interests, and promoting meaningful learning. That's easy enough to say, but it's often hard to do. It certainly helps when the school culture values personal commitment and academic achievement.

To make good things happen sometimes requires breaking down barriers to learning and moving students in the direction of academic success. It also requires opportunities for students to be active participants in their own learning.

Teachers need consistent societal and administrative support. Blaming teachers for most of the bad things that happen at school is counterproductive. After all, in one way or another, we are all involved in the education of children. So, all of us must do what we can; taking every opportunity to recognize the impact and positive influence that teachers can have on children's lives.

What does it take to become a good teacher? First of all, prospective teachers need to develop the intellectual tools learned in the arts and sciences. Equally important, they need to acquire the foundational skills that relate to strategic learning and pedagogical methods.

Even the best college graduates flounder if they start teaching without a thorough knowledge of the characteristics of effective instruction. Another factor in learning to be a successful teacher is early fieldwork experiences in different communities and in different school settings.

The research suggests that the most effective teachers are enthusiastic about their work and have high expectations for their students. This involves respecting students' creative potential and developing professionally appropriate personal connections. No one has to invade privacy to make good things happen. For teachers: it is probably best not to give out your phone number or home e-mail address because you need time away from schoolwork. Still, you need to recognize the fact that interpersonal relationships that develop over matters of content are at the heart of schooling.

Schools that involve teachers in participatory decision-making and collaborative strategies for addressing school problems will reinforce a teacher's commitment to the profession. Effective staff development programs have taken this into account as they address the challenge of sustaining the long-term commitment of dedicated and committed teachers. Through active participation in staff development programs, teachers can work in association with peers and "experts" to grow professionally and personally. Also, well-chosen university classes and professional conferences can certainly help.

By collaborating with a wide range of others to reach common goals, teachers are more likely to appreciate new instructional ideas and respect other professionals (Pahomov, 2014). Positive attitudes make a big difference. When teachers possess a positive attitude toward professional development and their own teaching, they will continue to be lifelong learners. It takes time to become a really good teacher.

SUMMARY, CONCLUSION, AND LOOKING AHEAD

Being socially responsible has a lot to do with contributing as much as possible to the local and global community. If conventional thinking gets in the way, try unconventional thinking. And remember, solutions to problems can often be found in unexpected places.

There are times when something that is part of the future gets replaced with something from the past. For example, in 2018, Silicon Valley's Elon Musk realized that to speed up production of the Model 3 Tesla (electric) car, he had to scrap a number of complicated robotic machines and replace them with human workers. In today's high-tech world, that was an unconventional approach—and it worked.

Unusual approaches, new ideas, and inventions are more than solving problems or making money on new products. They "shape our lives in unpredictable ways—and while they're solving a problem for someone, they're often making a problem for someone else. . . . [S]ometimes they slice through old constraints, and sometimes they weave entirely new patterns" (Harford, 2017, 11).

The problem isn't inventive technology. It's too few adults in children's lives to guide, monitor, and limit their use of that technology. In and out of school, learning and innovating require creative off-line thinking and face-to-face collaboration. At the early-childhood level, for example, children don't need screens as much as they need time for creative/collaborative play, hands-on exploration, and imaginative problem-solving.

Good citizenship requires more knowledgeable citizens. Few of us will reach the heights of creativity and invention or win a Nobel Prize. But it is possible—and necessary—for all of us to achieve a reasonable level of literacy and competency. Clearly, to participate in a democratic society, you have to keep up-to-date and have some idea of what's going on in the world.

Everyone has to play at least a supporting role if we are going to avoid dumbing down the future with shallow thinking. How? In part, by helping to increase the level of general understanding about the complexities surrounding the systems in which we are embedded. Or, as Benjamin Bratton put it in a TED talk, *more Copernicus, less Tony Robbins.*

> The future is not a result of choices among alternative paths offered by the present, but a place that is created—created first in mind and will, created next in activity. The future is not some place we are going to, but one we are creating. The paths to it are not to be found but created, and the activity of creating them changes both the maker and the destination.
>
> —John Schaar

NOTE

1. Valacich and Schneider (2017) define *artificial intelligence* as the science of using hardware, software, networks, and other technologies to make computers to reason and learn much as humans do, in addition to having sensing capacities.

REFERENCES

Agrawal, A., Gans, J., and Goldfarb, A. (2018). *Prediction Machines: The Simple Economics of Artificial Intelligence.* Cambridge, MA: Harvard Business Review Press.

Auerbach, D. (2018). *Bitwise: A Life in Code.* New York: Pantheon Books.

Beghetto, R., Kaufman, J., and Baer, J. (2015). *Teaching for Creativity in the Common Core Classroom.* New York: Teachers College Press.

Brynjolfsson, E., and McAfee, A. (2014). *The Second Machine Age: Work, Progress, and Prosperity in a Time of Brilliant Technologies.* New York: W. W. Norton.

Catmull, E. (2014). *Creativity, Inc.: Overcoming the Unseen Forces That Stand in the Way of Innovation.* New York: Random House.

Cummins, S. (2018). *Nurturing Informed Thinking: Reading, Talking, and Writing across Content-Area Sources.* Portsmouth, NH: Heinemann.

Dance, R., and Kaplan, T. (2018). *Thinking Together: 9 Beliefs for Building a Mathematical Community.* Portsmouth, NH: Heinemann.

Darling-Hammond, l. (2010). *The Flat World and Education: How America's Commitment to Equity Will Determine Our Future.* New York: Teachers College Press.

Dweck, C. S. (2012). *Mindset: How You Can Fulfill Your Potential.* London: Constable & Robinson.

Fox, M. (June 28, 2018). "She Knows How to Make an Exit. You're Reading It." *New York Times.* Retrieved from https://www.nytimes.com/2018/06/28/insider/obituary-writer-margalit-fox-retires.html.

Goodlad, J. (2016). *Romances with Schools: A Life of Education.* Reprint edition. Lanham, MD: Rowman & Littlefield.

Gruenert, S., and Whitaker, T. (2015). *School Culture Rewired: How to Define, Assess, and Transform It.* Alexandria, VA: Association for Supervision and Curriculum Development (ASCD).

Harford, T. (2017). *Fifty Inventions That Shaped the Modern Economy.* New York: Riverhead Books.

Hett, B. (2018). *The Death of Democracy.* New York: Henry Holt.

Johnson, S. (2014). *How We Got to Now: Six Innovations That Made the Modern World.* New York: Riverhead Books.

Marshall, J., and Donahue, D. (2014). *Art-Centered Learning across the Curriculum.* New York: Teachers College Press.

Opitz, M., and Ford, M. (2014). *Classroom Catalysts: 15 Efficient Practices That Accelerate Readers' Learning.* Portsmouth, NH: Heinemann.

Pahomov, L. (2014). *Authentic Learning in the Digital Age: Engaging Students through Inquiry.* Alexandria, VA: Association for Supervision and Curriculum Development (ASCD).

Pinker, S. (2014). *The Village Effect: How Face-to-Face Contact Can Make Us Healthier, Happier, and Smarter.* New York: Spiegel and Grau.

Safranski, R. (2017). *Goethe: Life as a Work of Art.* 1st edition. Translated by D. Dollenmayer. New York: Liveright.

Schaar, J. (1981). *Legitimacy in the Modern State.* New Brunswick, NJ: Transaction Publishers. Quote Retrieved from page 321, at https://books.google.com/books?id=aDE0WdWpIx0C&pg=PA321&dq=%22The+future+is+not+a+result+of+choices+among+alternative+paths+offered+by+the+present,+but+a+place+that+is+created%22&hl=en&sa=X&ved=0ahUKEwiDhdDwn57fAhXpxlkKHejTB44Q6AEIKDAA#v=onepage&q=%22The%20future%20is%20not%20a%20result%20of%20choices%20among%20alternative%20paths%20offered%20by%20the%20present%2C%20but%20a%20place%20that%20is%20created%22&f=false.

Valacich, J., and Schneider, C. (2017). *Information Systems Today: Managing in the Digital World.* 8th edition. White Plains, NY: Pearson Education.

Wagner, T. (2012). *Creating Innovators: The Making of Young People Who Will Change the World.* New York: Scribner.

Wetzell, R. F. (Editor). (2014). *Crime and Criminal Justice in Modern Germany (Studies in German History).* 1st edition. Brooklyn: Berghahn Books.

Chapter 1

Thinking Skills

Critical and Creative Thinking in a Technologically Intensive World

> Higher order thinking is the hallmark of successful learning at all levels—not only the most advanced.
>
> —Lauren Resnick

Educators have long realized that it takes more than knowledge of a subject to generate imaginative ideas and figure out clever solutions to problems. So, it is little wonder that today's standards-based classrooms' original thinking is valued more than ever.

In creatively supportive schools, the focus is on the development of thinking skills rather than telling students what to think.

Team-centered inquiry and imaginative problem-solving are viewed as central to the process of dealing with critical concepts and producing thoughtful results. But whatever the topic, unlocking the power of the imagination is viewed as key to sparking students' thinking and learning.

Encouraging original thinking within collaborative structures cuts across what to teach (*curriculum*) and how to teach (*instruction*). In many ways, digital technology is the biggest new kid on the block. In some ways, tech tools can help; in other ways, they may prove to be more damaging than originally thought.

All across the curriculum, critical and creative thinking are viewed as integral parts of today's elevated concept of subject-matter competency. Although there is a great deal of overlap, there are some differences in the various thought processes. Critical thinking, for example, has a lot to do with reasoning, criticism, logical analysis, and searching for supportive evidence. Creative thinking may be viewed as fluency, flexibility, originality, and elaboration—including unique expressions, novel approaches, and demonstrations of the ability to see things in imaginative and unusual ways.

The ability to effectively deal with the convergence of thinking and technological tools is one of the keys to educational and personal success. Take the always evolving Internet, virtual reality, and artificial intelligence as examples. They can help us extend our imaginations and connection to others. On the other hand, the same technology can become a distraction and get in the way of understanding the real world.

MEANINGFUL THINKING AND REASONING SKILLS

> Out of the questions of students come most of the creative ideas and discoveries.
>
> —Ellen Langer (1998)

The thinking of every child can be enhanced if the learning is *challenging*; this means that, in addition to providing proof of new facts and experiences, there must also be *group feedback*. Challenges can be as simple as presenting original content, including journal writing and hands-on materials. It's important to often change teaching strategies, to apply computer programs, have students work in groups, give students choice when doing projects, and take field trips.

Student feedback should not be overlooked. This builds confidence, reduces stress, and increases thinking's coping abilities. Students' peers are the greatest stimulus in the classroom. Collaborative groups provide students with a sense of feeling valued and cared for.

Thinking is a search for meaning, a purposeful quest for understanding and clarity. This journey often provides new points of view and solutions to problems. Thinking is intentional, purposeful, and deliberate. Individuals can be resourceful and adventuresome. Stimulating their students' critical and creative imagination is one of the most important teaching challenges teachers face today.

Thinking is built on personal experiences. It is also influenced by emotions, culture, home environment, and educational possibilities. Critical and creative thinking experiences express original ideas and solutions. They have a conscious and mental focus. The playful spirit of *creative thinking* can occur while daydreaming, fantasizing, or just having an idea while taking a hike along a trail. Creative thinking extends art and beauty as it reaches beyond the adequate to try elegant solutions.

Critical thinking is constructing meaning by observing, interpreting, analyzing, and manipulating information in response to a problem. Clarifying and solving problems, pondering alternatives, strategically planning, and analyzing the results are all activities that support critical thinking. *Creative thinking* is flowing, flexible, novel, and detailed. Creative thinking skills try to create novel expressions, unique conceptions, and original approaches. The mental power to see things in an unusual and imaginative way is linked with problem-solving and is part of critical thinking.

Critical thinking lessons might include having students analyze the hidden assumptions that produce meanings and different results of data information. Such intellectually demanding thinking helps students identify, clarify, and solve problems. The questions explored can be as general as this: "Can we be certain about the knowledge of this subject?" Questions can be as simple as "How was that done?" or "What does that mean?"

OVERLAPPING THINKING DOMAINS

Critical thinking often involves a persistent effort to examine information in light of reliable evidence. Along with carefully selected information, creative and critical reasoning can reinforce each other. Thinking transforms the information gathered by our digital devices and that information invites thinking (Howie, 2011).

Creative thinking tends to look at problems or situations from an imaginative perspective and to suggest fresh solutions. Students might be stimulated by methods ranging from brainstorming to the step-by-step processes of lateral thinking (O'Toole and Beckett, 2009).

Developing approaches that extend students' thinking skills across the curriculum is a strategy that is widely supported. Both the concept and the teaching strategies associated with thinking skills can be found in just about every curriculum. Better yet, most teachers are building at least some pieces of the thinking-skills puzzle into their day-to-day work.

TEACHING, LEARNING, AND THINKING HABITS

Understanding the relationship between theory, research, and practice is the foundation of pedagogical knowledge. The content standards build on this cycle to give a coherent, professionally defensible conception of how a subject can be framed for instruction. They do not put forward one best way to teach subject matter *or* thinking skills. Instead, all the standards projects have left room for interpretation.

A good place to start with students centers around real-life experiences that evoke personal meaning. To paraphrase the writer David Foster Wallace, it's important to see the most obvious and relevant realities around you, rather than getting too caught up in yourself. To make his point, he has looked to an old joke about two young fish out for a swim. They encounter an older fish who asks, " Morning, boys, how's the water?" As the younger fish swims on, one asks the other, "What the hell is water?"

Across all subjects, high-quality lessons often begin with real materials and make use of interactive learning in a way that allows students to explore the many dimensions of thoughtfulness, subject matter, and real-world applications. The basic idea is to help students form a new set of expectations and create a new sense of understanding. Asking the right questions certainly helps.

We all make sense of something by connecting to a set of personal, everyday experiences. So, good teachers have always connected educational goals to practical problem-solving and students' life experiences. This way, thinking skills are introduced into the curriculum so that students are intensely involved in reasoning, elaboration, forming hypotheses, and problem-solving. New ideas of literacy will have to move beyond disciplinary boundaries.

Using new methods for teaching mathematics, science, the arts, and language arts depends on reflective teachers. This means that both beginning and veteran teachers should take courses in learning math and science through inquiry and learn to apply the arts and language arts concepts within a context similar to the one they will arrange for their students. The result enlarges horizons and organizational possibilities.

Creative and critical thinking are natural human processes that can be extended by awareness and practice. Classroom instructional practice in the development of these skills might include the following:

1. Be accurate, clear, open-minded, and sensitive to others; defend a position.
2. Engage in difficult problems; extend the limits of your knowledge; find new ways to look at situations outside conventional boundaries; dare to imagine, innovate, and trust.
3. Listen with understanding when listening to other persons and try to understand their point of view. This is one of the highest forms of communication.
4. Be aware of your thinking: be sensitive to feedback, plan, and evaluate your actions. Take your time, remain calm, and think before you act.
5. Be attentive. Attention calls for skills such as observing, getting information, forming questions, and using inquiry.
6. Pose clear questions in easy-to-understand language.
7. Remain open to continuous learning.
8. Analyze and form hunches. Analysis is at the heart of critical and creative thinking. This includes recognizing and articulating attributes and focusing on details and structure, identifying relationships and patterns, and finding errors.

9. Use models and metaphors. Such higher-order thinking involves making comparisons, using metaphors, producing analogies, and providing explanations.
10. Gather data and assess and evaluate ideas. This establishes criteria or helps verify data.
11. Take risks. Elegant solutions oftentimes demand risk-taking and thinking independently.
12. Search out humor. This frees creativity and stimulates higher-level thinking skills. People who initiate humor are verbally playful when interacting with others; they also thrive on being able to laugh at situations and themselves.

Every inquiry, if explored with enthusiasm and with care, will use some of these core thinking skills (Keeley, 2018).

ILLUMINATING LEARNING BY TEACHING FOR THOUGHTFULNESS

The multidimensional search for meaning is made at least a little easier when there is a supportive group climate for generating questions and investigating possibilities. Thoughtful questions may also come into play after collaborative solutions are put forward.

Ask students to analyze the interesting questions they have asked and how they approached the problems they have dealt with. What about success, failure, and unexpected results?

As students examine how underlying assumptions influence interpretations, they can be pulled more deeply into a topic. This works best in small groups (Udvari-Solner and Kluth, 2018). By evaluating their findings on the basis of logic, students can generate a wide range of possibilities.

To have power over the story that dominates one's life in these technologically intensive times means having the power to retell it, deconstruct it, joke about it, and change it as times change. Without this power, it is more difficult to think and act on new thoughts and open the doors to deep thinking.

The core capacity of any subject includes more than knowledge acquisition. If you want students to come up with a better way to think and do something about it, *encourage them to find it*. The old view of teaching as the transmission of content has been expanded to include new intellectual tools and new ways of helping students thoughtfully construct knowledge on their own and with peers.

Teachers who invite thoughtfulness understand that knowledge is to be shared or developed rather than held by the authority. They arrange instruction so that children construct concepts and develop their thinking skills. As a result, everyone involved becomes an active constructor of knowledge and more capable of making thoughtful decisions in the future.

Recognizing the development of thinking skills is a good first step toward its application and assessment. Beyond specific teaching strategies, the climate of the classroom and the behavior of the teacher are very important. Teachers need to model critical thinking behaviors—setting the tone, atmosphere, and environment for learning. Being able to collaborate with other teachers can make a formative contribution to how the teacher might better see and construct individual classroom reality.

In collaborative problem-solving, teachers can help each other in the clarification of goals. They can also share the products of their joint imaginations. Thus, perceptions are changed, ideas flow, and practice can be meaningfully strengthened, deepened, and extended. Like their students, teachers can become active constructors of knowledge.

A curriculum that ignores the powerful ideas of its charges will miss many opportunities for illuminating knowledge. To teach content without regard for self-connected thinking prevents subject-matter knowledge from being transformed in the student's mind. If the curriculum is to be viewed as *enhancing being* and *opening to the unfamiliar*—rather than merely imparting knowledge and skills—then, reasoned and shared decision-making is part of the process.

Encouraging fresh ideas or opposing views is often difficult for administrators. All of us need the occasional push or encouragement to get out of a rut. Breaking out of established patterns can be done collectively or individually. But it takes those most directly involved to make it happen.

JOURNEYS OF DISCOVERY

> When human beings create and share experiences designed to delight and amaze, they often end up transforming society in more dramatic ways than people focused on more utilitarian concerns.
>
> —Steven Johnson (2017, 11)

Imaginative play is one way to open doors to innovative thinking, collaborative inquiry, and problem-solving. It is important for all of us to develop our own ways of reflection and becoming students of our own thinking.

Can teachers make a difference? Absolutely. Nothing is more powerful than connecting willing teachers with innovative methods and materials so that they can build learning environments that are sensitive to students' growing abilities to think for themselves.

Learning by example certainly helps generate student creativity. So, when a teacher decides to participate with students in learning to think creatively and critically on a daily basis, they are more likely to nourish human possibilities.

The Common Core suggests that there are certain student thinking skills that are needed across the curriculum. By promoting thoughtful learning across the full spectrum of personalities, cultures, and ways of knowing, teachers can make a tremendous difference and perform a unique service for the future.

When the ideal and the actual are linked, the result can produce a dynamic, productive, and resilient form of learning. What we know is that teaching and thinking are increasingly being put into practice in model classrooms and schools. These exemplary programs recognize that powerful inquiry can help students make personal and group discoveries that change thinking. Good thinking skills can turn an unexamined belief into a reasoned one.

By nurturing informed thinking and awareness, we can all learn how to actively apply knowledge, solve problems, and enhance conceptual understanding across social boundaries. As children use reason and logic to change their own theories and beliefs, they grow in ways that are personally meaningful. Understanding the essence of contradictory points of view means understanding some of the universal truths that speak to everyone. A diversity of new voices can add vigor to understanding the world and our place in it.

As the Common Core and each of the standards projects point out in their own way, notions about how students learn a subject need to be pluralized. Almost any important concept can be approached from multiple entry points—emphasizing understanding and making meaningful interdisciplinary connections.

As students learn about the perspectives of other cultures—including varying social and historical backgrounds—they can explore where stereotypes come from. With a little homework, each student can design a large, graphic family tree to share. This way, each student's cultural background can be a valuable tool for learning about themselves, their own cultural background, and how their communities connect to other communities around the world.

With the globalization of media and business, it becomes ever more important to see how events in the United States affect people in other countries—and vice versa. Gaining a global perspective means developing a more integrative understanding of the human community and overlapping cultural experiences. As teachers learn to thoughtfully view the world from multiple perspectives, the way is cleared for them to become more sensitive to variation and more capable of reaching diverse learners.

TOWARD A NEW VISION OF INTELLIGENCE—MULTIPLE INTELLIGENCES (MI)

Learning has a lot to do with finding your own gifts (Armstrong, 2017). To make learning more accessible to children means respecting multiple ways of making meaning.

The brain has a multiplicity of functions and voices that speak independently and distinctly for different individuals. Howard Gardner's framework for multiple entry points to knowledge has made an impact on the content standards. There are many differences, but each set of content standards is built on a belief in the uniqueness of each child and the view that this can be fused with a commitment to achieving worthwhile goals.

Lessons built on Gardner and Davis's (2013) ideas proved helpful providing alternative paths for learning.

Multiple Intelligences

1. Linguistic intelligence: the capacity to use language to express ideas, excite, convince, and convey information. Speaking, writing, and reading.
2. Logical-mathematical intelligence: the ability to explore patterns and relationships by manipulating objects or symbols in an orderly manner.
3. Musical intelligence: the capacity to think in music, the ability to perform, compose, or enjoy a musical piece. Rhythm, beat, tune, melody, and singing.
4. Spatial intelligence: the ability to understand and mentally manipulate a form or object in a visual or spatial display. Maps, drawings, and media.
5. Bodily-kinesthetic intelligence: the ability to use motor skills in sports, performing arts, or art productions, particularly dance or acting.
6. Interpersonal intelligence: the ability to work in groups. Interacting, sharing, leading, following, and reaching out to others.
7. Intrapersonal intelligence: the ability to understand one's inner feelings, dreams, and ideas. Introspection, meditation, reflection, and self-assessment.
8. Naturalist intelligence: the ability to discriminate among living things (plants, animals) as well as a sensitivity to the natural world.

Gardner and Davis define intelligence as the ability to solve problems, generate new problems, and do things that are valued within one's own culture. MI theory suggests that these eight intelligences work together in complex ways. Most people can develop an adequate level of competency in all of them. And there are many ways to be "intelligent" within each category.

Still, there is general agreement on a central point: *intelligence is not a single capacity that every human being possesses to a greater or lesser extent.*

There *are* multiple ways of knowing and learning. And whether or not we subscribe to MI theory, methods of instruction should reflect different ways of knowing.

Suggestions for Using Multiple Intelligence Activities

1. Put Multiple Intelligence theory into action. Some possibilities include those listed in the table on the next page (Table 1.1).
2. Build on students' interests. When students do research either individually or with a group, allow them to choose a project that appeals to them. Students should also choose the best way for communicating their understanding of the topic. In this way, students discover more about their interests, concerns, learning styles, and intelligences.
3. Plan interesting lessons. There are many ways to plan interesting lessons.

Lesson Planning

1) Set the tone of the lesson: Focus student attention and relate the lesson to what students have done before. Stimulate interest.
2) Present the objectives and purpose of the lesson: What are students supposed to learn? Why is it important?
3) Provide background information: What information is available? Resources such as books, journals, videos, pictures, maps, charts, teacher lectures, class discussions, or seat work should be listed.
4) Define procedures: What are students supposed to do? Instructions given should include examples and demonstrations as well as working directions.
5) Monitor students' understanding: During the lesson, the teacher should check students' understanding and adjust the lesson if necessary. Teachers should invite questions and ask for clarification. A continuous feedback process should be in place.
6) Provide guided practice experiences: Students should have a chance to use the new knowledge presented under direct teacher supervision.
7) It is equally important that students get opportunities for independent practice where students can use their new knowledge and skills.
8) Evaluating students' work is necessary to show that students have demonstrated an understanding of significant concepts. Remember, paper-and-pencil tests do not adequately measure students' critical and creative thinking.

Table 1.1 Multiple Intelligence Activities

linguistic intelligence	musical intelligence	logical-mathematical intelligence	spatial intelligence	bodily-kinesthetic intelligence	interpersonal intelligence	naturalist intelligence	intrapersonal intelligence
write an article	sing a rap song	design and conduct an experiment	illustrate, draw, paint, sketch	use creative movement	conduct a meeting	prepare an observation notebook	write a journal entry
develop a newscast	give a musical presentation	describe patterns	create a slide show, videotape, chart, map, or graph	design task or puzzle cards	participate in a service project	describe changes in the environment	describe one of your values
make a plan	explain music similarities	make up analogies to explain	create a piece of art	build or construct something	teach someone	care for pets, wildlife, gardens, or parks	assess your work
describe a procedure	play a musical instrument	solve a problem		bring hands-on materials to demonstrate	use technology to explain	use binoculars, telescopes, or microscopes	set and pursue a goal
write a letter	demonstrate rhythmic patterns			use the body to persuade, console, or support others	advise a friend or fictional character	photograph natural objects	reflect on or act out emotions
conduct an interview write a play interpret a text or piece of writing							

4. Portfolios represent the cutting edge of more authentic and meaningful assessment. They are powerful assessment tools that require students to select, collect, and reflect on what they are creating and accomplishing.

A SAMPLE MULTIPLE INTELLIGENCE LESSON PLAN

Lesson Title: How Intelligence Cells Work

Students should develop understandings of personal health, changes in environments, and local challenges in science and technology. The human body and the brain are fascinating areas of study. The brain, like the rest of the body, is composed of cells; but brain cells are different from other cells.

This lesson focuses on the science standards of inquiry, life science, science and technology, and personal and social perspectives.

Lesson Goals

The basic goal is to provide a dynamic experience with each of the eight "intelligences." The next step is using graphic imagery to map out the selected intelligences on construction paper.

Procedures

1. Divide the class into groups. Assign each group an intelligence.
2. Allow students time to prepare an activity that addresses their intelligence. Each small group will give a three-minute presentation (with a large map) to the entire class.

Objective

To introduce students to the terminology of intelligence and how intelligence functions—specifically, the function of intelligence cells.

Grade level

With modifications, K–8.

Material

paper, pens, markers

Intelligence (Thinking) "Recipe"

5 cups of instant potato flakes, 5 cups of hot water, 2 cups of sand
1 gallon Ziploc bag

Combine all ingredients in the Ziploc bag; mix thoroughly. It should weigh about 3 pounds and have the consistency of a real part of our thinking process in our head.

Background information

No one understands exactly how thinking works. But scientists know the answer lies within the billions of tiny cells, called *nerve cells*, which fire up the thinking process. All the body's feelings and thoughts are caused by the electrical and chemical signals passing from one neuron to the next. A nerve cell looks like a tiny octopus, but with many more tentacles (some have several thousand). Nerve cells carry signals throughout our thinking process that allows us to move, hear, see, taste, smell, remember, feel, and think.

Procedure

1. Make a model of the thinking process to show to the class. The teacher displays our thinking process and says, "The smell of a flower, the memory of a walk in the park, the pain of stepping on a nail—these experiences are made possible by the three pounds of tissue in our heads— "OUR THINKING PROCESS."
2. Show a picture of the nerve cell and mention its various parts.
3. Have students label the parts of the cell and color if desired.

Activity 1 Message Transmission: *Explaining how thinking (nerve) cells work*

A message traveling in the nervous system of our intelligence can go 200 miles per hour (mph). These signals are transmitted from intelligence cell to cell across a connection. To understand this system, have students act out the thinking cell process.

1. Instruct students to get into groups of five. Each group should choose a group leader.
2. Direct students to stand up and form a circle. Each person is going to be an intelligence or a thinking cell. Students should be an arm's length away from the next person.
3. When the group leader says "Go," have one person from the group start the signal transmission by slapping the hand of the adjacent person. The second person then slaps the hand of the next, and so on until the signal goes all the way around the circle and the transmission is complete.

Explanation

The hand that receives the slap is the *branching part* of the nerve cell. The middle part of the student's body is the *cell body*. The arm that gives the slap to the next person is the *nerve cell*, and the hand that gives the slap is the *nerve cell terminal*. In between the hands of two people is the *nerve cell connector*.

Inquiry Questions

As the activity progresses, questions will arise: What are parts of a thinking nerve cell? A tiny nerve cell is one of billions that make up the thinking process. A nerve cell has three basic parts: the *nerve cell,* the *cell body*, and the *nerve cell connector.*

Have students make a simple model by using their hand and spreading their fingers wide. The hand represents the "cell body," the fingers represent the part that brings information to the cell body, and the arm represents the "cell connector," which takes information away from the cell body.

Just as students wiggle their fingers, the nerve cells are constantly moving as they seek information. If an intelligence cell needs to send a message to another cell, the message is sent out through the nerve cell. The wrist and forearm represent the cell body. When a cell sends information down its cell body to communicate with another nerve cell, it never actually touches the other cell. The message goes from the nerve cell of the sending cell to another nerve cell by "swimming" through the space called the *cell connector.*

Neuroscientists define *learning* as *two nerve cells communicating with each other.* They say that nerve cells have "learned" when one cell sends a message to another cell.

MULTIPLE INTELLIGENCES LEARNING ACTIVITIES

Linguistic: writing a reflection about the activity, researching how a nerve cell works, keeping a study journal about how nerve cells work
Bodily/Kinesthetic: move like a nerve cell
Group Drama: cell signal transmission
Visual/Spatial: mapping the connections of our thinking process (connect the dots)
Musical: singing songs about nerve cells; tapping out rhythms to the song "Because I Can Think"
Naturalist: describing changes in your thinking environment illustrating a thinking connection

Interpersonal: participate in (act out) a group signal, nerve-cell transmission observing/recording
Intrapersonal: reflecting on thinking; keeping a journal of how the brain works
Mathematical/Logical: calculating nerve cell connections

Evaluation

Each group will write a reflection on the activity. Journal reflections should tell what they learned about thinking and how that helps them understand how the thinking process works. How might this theory influence a STEM lesson?

STEM ACTIVITIES: KINDERGARTEN AND FIRST GRADE

The research suggests that students do better with science, technology, engineering, and mathematics (STEM) in the upper grades if they start with STEM activities early on (Hierck and Freese, 2018). STEM content can be learned through investigation, play, and focused intentional teaching. Sometimes, it's as simple as asking better questions.

Example of a STEM Inquiry Approach to Sand Tables

Materials

sand tables that are found in many K–1 classrooms
marbles, rulers, cups, and small boxes

Procedure

Children gather around a sand table and explore the sand and let some of the grains run through their fingers. The teacher, then, provides some props: marbles, rulers, cups, and small boxes.

Possible Directions and Questions

Be sure that the students have time after each direction or question.

- "Get a partner and a set of tools."
- "How can you make the marbles roll on the sand?"
- "What do you think makes your marble slow down or stop as you roll it across the sand?"

- See if the students come up with using the ruler. If not, give a few hints.
- "Why did it go faster on a ruler than in the sand?"

After the lesson is over, bring the whole class back together and briefly discuss the results.

There is general agreement that basic technologies like rulers, blocks, and balance scales are fine for younger children. But when it comes to *digital* technology, the research is mixed. Still, you should know that there are online activities for the early grades.

An example of a website for student STEM activities at the K–3 level, *ScratchJr* was developed by MIT and *Playful Inventions* to bring STEM concepts to the early childhood level. A useful book to use in designing STEM activities is by editors M. Honey and D. Kanter: *Design, Make, Play: Growing the Next Generation of STEM Innovators* (2013).

STEM ACTIVITIES: MIDDLE GRADES AND UP

The National Science Teachers Association (NSTA) suggests integrating the intellectual tools of STEM with subjects ranging from the language arts to social studies and the arts.

One example is "STEM with an Environmental Focus." NSTA points to schools using an "E-STEM" procedure that emphasizes environmental issues and sustainability. This is one of their examples: a class where students collaboratively explore how the ecospheres (biosphere, lithosphere, hydrosphere, and anthroposphere) are made or modified by humans. NSTA reports that such an approach amplifies "student innovation and curiosity about real world situations."

A Lesson Example: Thinking about Rebooting the World

Astrobiologist Lewis Dartnell suggests extending classroom STEM activities and suggested thought experiments designed by the Nobel prize–winning physicist Richard Feynman. This lesson example builds on these scientists' work exploring how human knowledge is collectively distributed across the population.

This activity involves having students come up with a starter kit for rebuilding civilization after an extinction event—such as an asteroid hitting earth, a nuclear war, or an Ebola-like virus spreading around the world.

Question

If all stored knowledge were destroyed, what key principles of STEM would be necessary to rebuild our world from scratch? Ask the following question: *If you could pass along just one STEM idea that would be key to moving toward a better future what would it be? . . . One sentence only.* After a little whole class explanation, the students can write their sentence.

Next, students work together in small groups where they share their idea. Have students write two or three of the original ideas in their list.

Mix the nine, ten, or eleven listed topics so that they are *not* written in order of "importance." After the small group has finished their final list of topics, have group members rank the topics on their list in terms of importance.

Directions for a group of three or four

Discuss and decide what the rank order of importance should be. [The most important should be listed first and the least important last.] After each group has their done their ranking, the whole class discusses the group choices. The following is a sample list:

- *Transform base substances:* information about how to transform things like clay and iron into brick and steel.
- *Germ theory:* Contagious diseases are caused by small organisms that you can't see. What to do? Drinking water, for example, can be disinfected with diluted bleach; hands can be cleaned with soap.
- *Scientific method:* a great knowledge-generating approach for things in the past and yet to be discovered. It's a method of research in which a problem is identified, observations made, data gathered, and a hypothesis is tested.
- *Atomic theory:* All things are made of atoms. These little particles constantly move around—attracting each other when they are some distance apart and repelling each other when tightly squeezed together.
- *Mathematical algorithm:* used in mathematics and computer science as a step-by-step procedure for solving computational problems. It's like a cooking recipe for mathematical calculation and problem-solving.
- *Agriculture and the ability to grow and stockpile food:* Cereal crops like rice, wheat, and maize have sustained civilizations for thousands of years.
- *Engineering:* the application of scientific and mathematical principles to the design and solution of practical problems; knowing how to design and build structures like bridges, buildings, and roads.

After the group choices are made, bring the whole class back together and have each group present and defend their list.

A 2014 book on the subject is *The Knowledge: How to Rebuild Our World from Scratch* by Lewis Dartnell.

CONTROLLING OUR TECH TOOLS

Throughout human history, the tools of creative production have been in the hands of a chosen few. Now, *the digital revolution has put sophisticated tools within reach of almost everyone. But even the most accessible of them are worthless without the ideas and expertise of those who use them* (Resnick, 2017).

Creative thinking and thoughtful problem-solving go hand in hand in today's tech world. Clearly, it is more important than ever to have students asking interesting questions and engaging in authentic inquiry.

The difference between whether or not new media is useful or useless has a lot to do with whether or not technology is treated with a measure of wisdom. Opening up new dimensions is one thing; encouraging escapism is quite another.

Our devices are becoming an ever more powerful force for contracting *and* expanding the human potential. They affect every aspect of our life and require both old and new ways of thinking. One thing that you can be certain of—users of digital media need to have greater control over how their time and personal information are being used (Heimans and Timms, 2018).

A good metaphor for the learning skills required in the digital age might be found in the difference between a sponge and panning for gold. The sponge approach emphasizes fairly passive knowledge acquisition; the panning for gold approach stresses active interaction with knowledge as it is being acquired. When used in an informed and disciplined way, new media can open up the possibility of imaginative solutions and alternative paths to the future. Unfortunately, they can also be a major distraction and a colossal waste of time.

SETTING LIMITS

How should limits be set up around technologies when the long-term effects on critical and creative thinking have yet to be figured out?

The ability to think and reflect can get lost in a world of addictive phones, exploitive websites, and online scams. Even knowledgeable adults often find it difficult to protect privacy and avoid predators in the midst of online advertising, personal temptations, and overuse. Unfortunately, it is even more dangerous for young users—especially when they are targeted as a prime audience.

Before children accidentally disappear into their screens, they need to learn how to sort through the digital clutter. Teachers, parents, tech companies, and young people themselves are key to getting the right off-line–online balance. So grown-ups—turn the phone off and have an occasional face-to-face conversation.

We have always known that new technologies produce problems—it's just that it's hard to figure out in advance what those problems will be. Today's digital media, for example, have helped push thinking from a relatively calm and focused process into something that needs to take in and dole out information in short, disjointed, and overlapping bursts (Carr, 2010).

Our mobile Internet age has produced some *digital natives* who expect continuous connection and feel uneasy when expected to think in the few moments they are truly alone (Turkle, 2011). It is little wonder that many students today have more trouble concentrating or thinking intelligently when they spend so much time skimming one surface after another to the point of distraction.

The educational path that was developed in the early twentieth century stressed the behaviorist view of learning. What could be seen was behavior. Learning was looked upon as changed behavior. Lacking the knowledge of what goes on inside the thinking process, the behaviorists measured behaviors and learned to alter them with behavioral reinforcement (rewarding positive behaviors and punishing negative actions).

As behaviorism faded from the scene, cognitive science moved in to fill the void. Most educators welcomed the change. *Cognition* may be thought of as the process of knowing. Cognitive science included a wide range of human activities: critical thinking, making judgments, decision-making, and creative thinking.

Cognitive science has moved the ball in the direction of *constructivism*. The basic idea being that the thinking individual does not simply take in knowledge, but is actively involved in constructing that knowledge.

In the classroom, good teaching encourages students to actively construct knowledge; within lessons that have clear goals, allow for guided practice, and provide ongoing assessment. Also, leaving some room for student choice amplifies critical and creative thinking.

TECHNOLOGY AS AN EXTENSION OF THE HUMAN MIND

We are living through the largest increases in expressive/thinking capacity in human history. Every new technology or idea has unexpected consequences for thinking. Still, it's hard to get specific about what they are going to be.

New machines generate new questions. For example: "How many human jobs will be taken over by robots or machines run by artificial intelligence?"

Although today's digital devices and apps can expand the traits found in the human mind, neither one of them has taken on a completely autonomous dimension.

Artificial intelligence (AI) is a good example of a technology that has finally reached a stage where it can extend human thinking and intelligence. Now, AI can act as a supplement to human thought in areas ranging from medical problems to making suggestions about what book we may want to read. When combined with other technologies like the Internet, the arena in which thinking resonates is vastly expanded.

Here are examples of AI applications that can assist learning:

- *Thinkster Math* is a math tutoring app that figures out a student's learning style and the best way for them to approach mathematical problem-solving. A digital tutor analyzes a student's mental processes in dealing with problems that are presented to him or her. The tutor points out what they have done right and what they have done wrong. Video and print are, then, used to provide feedback so that the student can better understand their strengths and weaknesses, and the concepts being studied.
- *Google Docs* has a new grammar check feature that checks and corrects the grammar of written works. The name of the app may change, but it has proven very useful for students with limited English language skills. We don't use a grammar checker because what's available now takes a little life out of what's being written.
- "Brainly" is a social media site where learners can go beyond standard subject-matter questions to figure out how to design important questions on their own. It is designed as a social media site that connects students to digital teachers and other students (real ones) who are dealing with some of the same academic work (Lynch, August 22, 2017).

There are times when both teachers and students can use powerful intellectual and tech tools to move from what *is* to what *might be*. The most important thing is to make sure that learners are becoming active doers rather than passive recipients of information. Making imaginative connections, curiosity, risk-taking, inquiry, and problem-solving are all essential parts of today's most up-to-date learning environments (Taddei and Budhai, 2017). Like education in general, human thinking is fluid and organic.

INTELLECTUAL TOOLS OF THE FUTURE

Since it is so difficult to figure out what knowledge will be crucial to students in the future, it makes sense to pay more attention to the *intellectual tools* that will be required *in any future*. This suggests focusing on how models of critical thought can be used differently, at different times, and in different situations. The idea is to put more emphasis on concepts with high generalizability—like collaborative problem-solving, reflection, perceptive thinking, self-direction, and the motivation needed for lifelong learning.

A more thoughtful and personalized brand of learning is a key to building bridges across the curriculum. The same can be said for active learner-centered collaboration. As far as STEM-influenced instruction is concerned, making subject-matter connections to real-world concerns is another part of the package (Weld, 2017).

Information isn't a substitute for thinking. But information and thinking are not antithetical. At higher levels, thinking requires quickly sorting through a wealth of information to be effective. There will never be enough time to teach all the information that we feel is useful. But time must be taken to be sure that student thinking can transform knowledge in ways that make it transferable to the outside world.

When there is time for inquiry and reflection, covering less can actually help students learn more deeply. A classroom where thinking is highly valued is most likely to develop learners who take creative and critical thinking seriously.

Within this context, the following thinking skills can be taught directly:

- generating multiple ideas about a topic
- figuring out meaning from context
- understanding analogy
- detecting reasoning fallacies

Topical knowledge (content facts), procedural knowledge (how to study and learn), and self-knowledge are all part of critical thinking. All of these thinking skills are learned through interaction with the environment, the media, peers, and the school curriculum. Some students pick it up naturally, while others learn reasoning skills with difficulty.

PROVIDING ACCESS TO THE THOUGHTFUL LIFE

Children can demonstrate what their reasoning ability is in a number of ways: think–out louds, videos, performances, photo-collages, stories for the newspaper, websites on the Internet, or multimedia projects that can be shared with other students and members of the community.

We are already seeing glimmers of a computer-based media that are broadly expressive and capable of capturing many aspects of human consciousness.

As the twenty-first century progresses, the whole spectrum of expression is being altered. Now, many people form an opinion and look for facts to support it. In science today—and in the general population yesterday—facts are gathered before forming opinions.

Communication and information technology sometimes complement and sometimes supersede previous media. Still, the basic learning process and the essence of any curriculum will continue to involve ways of engaging students in thought that matters and sharing what they find—information/knowledge/wisdom. Wisdom is important because it gives you the power to change the shape of ideas.

Helping students learn the truth of others can make it possible for them to discover their own. Feeling and meaning can be turned inside out as students learn how to construct their own knowledge and absorb new learning experiences in ways that make sense to them. This extends to anticipating and exploring (from many angles) the depths that await us under the surface of things, whatever those things may be.

There is always the possibility of undermining our chances for imaginative thinking by getting caught up in what David Whyte calls "the eddies and swirls of everyday existence."

I turned my head for a moment and it became my life.

—David Whyte

SUMMARY, CONCLUSION, AND LOOKING AHEAD

The creation of something new often goes against traditional approaches and the authority of the present. Also, creative and critical thinking are by their very nature subversive and outside the specified lines of behavior. Still, there is nothing wrong with setting specific goals and requiring thoughtful products.

Thoughtful teaching practice, at its best, is enriched by theory and flexible enough to be altered in the light of practice. Research also plays a clarifying role in this complementary process. Subjects across the curriculum are grounded in all of the above in an effort to move students along the path to becoming independent learners.

Integrating social skills with language, literacy, and the other core subjects is proving to be the most effective way to deal with high-tech possibilities like automation, robotics, and artificial intelligence. Whether it is high-tech or low-tech, as far as creative and critical thinking are concerned, it is best to weave these skills into the entire classroom fabric in a way that connects across the curriculum.

Like it or not, learning in a socially connected world changes both *what* and *how* we think. So, it is little wonder that the social side of expanding our reality (thinking) has an increasingly crucial role to play in schooling.

Digital and information literacy matter, but it isn't necessary for everyone to learn a specific skill like coding. Although a few schools have chosen to teach computer coding skills, robots are coming along that will do a lot of the coding work for us.

Particularly at the primary level, computers can get in the way of developing thinking skills. In spite of pressures to move in other directions, it would be wise for schools to pay close attention to fundamental human skills like face-to-face communication, teamwork, and critical/creative thinking.

Remember, thinking skills are part of a process that builds on previous experience in a way that helps learners build understanding and knowledge (Rhoton, 2018).

Tomorrow's schools are bound to view thinking skills as essential to the core curriculum. Along the path to reason and thoughtfulness, it is best to plan instruction so that you can tune in to what students are doing and build on their interests. Even in the middle of a lesson, it is always possible to make adjustments that reflect patterns of thinking.

Although history usually doesn't repeat itself, it does present some useful lessons. So, don't forget to look in the rearview mirror. Here's an educational example: It has always been viewed as helpful when students have had access to real-world, applied learning experiences. And the best route for personal and group success is—and has always been—to optimize the available possibilities.

Learning to ride wave after wave of unpredictable change requires a curious, inquisitive, and persistent mind. To paraphrase Pasteur, *change has always favored the prepared mind.*

> There is no doubt that creativity is the most important human resource of all. Without creativity there would be no progress and we would be forever repeating the same patterns.
>
> —Edward de Bono

REFERENCES

Armstrong, T. (2017). *Multiple Intelligences in the Classroom.* 4th edition. Alexandria, VA: Association for Supervision and Curriculum Development (ASCD).

Carr, N. (2010). *The Shallows: What the Internet Is Doing to Our Brains.* New York: W. W. Norton.

Dartnell, L. (2014). *The Knowledge: How to Rebuild Our World from Scratch.* New York: Penguin Group (USA).

Division of Behavioral and Social Sciences and Education (Author), Commission on Behavioral and Social Sciences and Education (1987). *Education and Learning to Think.* Washington, DC: National Academies Press.

Gardner, H., and Katie Davis. (2013). *The App Generation: How Today's Youth Navigate Identity, Intimacy, and Imagination in a Digital World.* New Haven, CT: Yale University Press.

Heimans, J., and Timms, H. (2018). *New Power: How Power Works in a Hyperconnected World—and How to Make It Work for You.* New York: Doubleday.

Hierck, T., and Freese, A. (2018). *Assessing Unstoppable Learning.* Bloomington, IN: Solution Tree Press.

Honey, M., and Kanter, D. E. (Editors). (2013). *Design, Make, Play: Growing the Next Generation of STEM Innovators.* New York: Routledge.

Howie, D. (2011). *Teaching Students Thinking Skills and Strategies.* London: Jessica Kingsley Publishers.

Johnson, S. (2017). *Wonderland: How Play Made the Modern World.* New York: Riverhead Books.

Keeley, P. (2018). *Uncovering Student Ideas in Science.* 2nd edition. Arlington, VA: National Science Teachers Association (NSTA) Press.

Langer, E. J. (1998). *The Power of Mindful Learning.* Cambridge, MA: Da Capo Press.

Lynch, M. (August 22, 2017). "5 Examples of Artificial Intelligence in the Classroom." The Tech Edvocate. https://www.thetechedvocate.org/5-examples-artificial-intelligence-classroom/.

O'Toole, J., and Beckett, D. (2009). *Educational Research: Creative Thinking and Doing.* Oxford: Oxford University Press.

Resnick, M. (2017). *Lifelong Kindergarten: Cultivating Creativity through Projects, Passion, Peers, and Play.* Cambridge, MA: MIT Press.

Rhoton, J. (Editor). (2018). *Preparing Teachers for Three-Dimensional Instruction.* Arlington, VA: National Science Teachers Association (NSTA) Press.

Taddei, L. M., and Budhai, S. S. (2017). *Nurturing Young Innovators: Cultivating Creativity in the Classroom, Home, and Community.* Portland, OR: International Society for Technology in Education.

Turkle, S. (2011). *Alone Together: Why We Expect More from Technology and Less from Each Other.* New York: Basic Books.

Udvari-Solner, A., and Kluth, P. (2018). *Joyful Learning: Active and Collaborative Strategies for Inclusive Classrooms.* 2nd edition. Thousand Oaks, CA: Corwin.

Weld, J. (2017). *Creating a STEM Culture for Teaching and Learning.* Arlington VA: National Science Teachers Association.

Whyte, D. (1994). *The Heart Aroused.* New York: Currency/Doubleday.

Chapter 2

Collaborative Learning

Teamwork and Social Learning Strategies

There is much we do alone. But together we can do more.

—L. S. Vygotsky

Having students work together with a partner or in a small group is a good way to actively construct meaning across the curriculum. In one form or another, collaborative learning is one of the more important instructional tools to gain wide acceptance since the 1960s. So, it's little wonder that the content standards and Common Core projects recommend certain elements of collaborative or cooperative learning for reaching a diverse group of students.

Collaborative learning builds on what we know about the social nature of learning and how students construct knowledge. The basic idea is to promote active learning in ways not possible with highly competitive or individualized learning models. In the classroom, the various approaches to social learning all suggest cooperative contact among a small group of peers. Although there is individual accountability, the group either sinks or swims together.

Teamwork is viewed as one of the keys to accelerating students' imaginative development and academic achievement. In an interactive learning environment, students serve as learning resources for each other. This requires students' talking to each other, checking each other's work, discussing concepts, and making collaborative decisions (Strebe, 2018).

Innovation is a good example of an imaginative process that relies on teamwork and close collaboration. Sometimes, this means small steps; at other times, it involves quick, fundamental change. Knowledge is a major innovation generator. Whether it is fast, slow, or incremental, transformational change can be awkward. Still, getting smarter about trying new ideas and updating old assumptions is a potent engine for social and economic progress.

Don't underestimate the role of social culture and chance; a random roll of the dice has always had a lot to do with outcomes. Archaeologists have long suspected that the emergence of modern humans involved major upgrades in social interaction, communication, and technology. Creating a critical mass for new ideas and progress has always involved changes in *how* and *what* people thought—with cultural evolution often playing a bigger role than genetic evolution (Heyes, 2018).

In some ways, today's online social networking is a useful method, product, and service. But is it an example of useful innovation? Some say yes, others see platforms like Facebook as getting in the way of authentic group interaction. Whether you like it or not, there are a lot of distracting swamp creatures out there, so use digital technology with caution.

When it comes to collaboration and knowledge building in the physical classroom, the teacher organizes major parts of the curriculum around tasks, problems, and projects so that students can work together in small, mixed-ability groups.

COLLABORATION AS AN APPROACH TO LEARNING

Although the terms collaborative learning and cooperative learning are sometimes used interchangeably, there are differences. In cooperative learning, for example, there tends to be a greater emphasis on positive interdependence and individual accountability (Johnson et al., 2008).

With cooperative learning, student roles may be more defined: reader, checker, encourager, and recorder (in a group of four).

- The *reader* reads and explains the problem to the group.
- The *checker* makes sure that everyone understands the problem.
- The *encourager* takes an extra special responsibility for keeping the work interesting and on task.
- The *recorder* writes down how group members figure out a problem and reports the solution.

If there are three students in a group, one of them takes on two roles. Everyone actively participates. At the end of the assignment or project, students evaluate their work and how well each of them played their role.

To avoid getting bogged down with definitions, we pay more attention to the collaborative side of the equation. No matter how the group work is arranged, it is our belief that students do better when they are immersed in learning through interaction with others. Also, they are more likely to remember something they discover together rather than simply collecting information from the passive acceptance of teacher presentations (Stronge, 2018). Getting the right balance is the key to good teaching. Collaborative learning may be thought of as a classroom technique and a personal teaching philosophy. Although the small group often "sinks or swims" together, individuals are held accountable and students receive feedback from peers and the teacher. Teamwork matters, but so does respect for individual group members' abilities and contributions. As learners learn to collaborate, group work can enhance everyone's knowledge, proficiency, and enjoyment.

To get ready for classwork, students can be taught to quickly arrange the room based on the activity; it might be a team of three or four, or a pair structure. Sometimes, it is best to arrange the room before students come in.

Group responsibility and individual accountability are important factors in a collaborative classroom. Also, there is a sharing of authority and acceptance of responsibility among group members for the group actions. The end result depends on consensus building and cooperation among group members.

The content standards in mathematics, science, language arts, and other subjects recommend having students collaborate as they go about doing some of their schoolwork. In addition, student talk is stressed as a way for working things out among group members (National Research Council, 1996). What's missing from the standards are specific activities and organizational techniques for making collaborative groups work in the classroom.

Collaborative learning builds on what teachers know about how students construct knowledge, promoting active learning in a way not possible with competitive or with individualized learning. In a cooperative classroom, the teacher organizes major parts of the curriculum around tasks, problems, and projects that students can work through in small mixed-ability groups. Lessons can be designed around active learning teams in a way that helps students combine energies and reach toward a common goal.

Social skills, like interpersonal communication, group interaction, and conflict resolution, are developed as the collaborative learning process goes along. Students soon get the idea that if someone else does well, you can do well just the same. After each lesson, the learning group examines what they did well and what they might be able to do better (social processing).

RESEARCH-BASED AND STANDARDS-DRIVEN APPROACH

For decades, research has suggested that collaborative learning has positive effects (Baker et al., 2013). A selection of findings follows:

- *Collaborative learning can motivate students who are having difficulties with various subjects.* Students talk and work together on a project or problem and experience the fun of sharing ideas and information.
- *Classroom interaction with others causes students to make significant learning gains compared to students in traditional settings.*
- *Collaborative learning may help encourage active listening for disinterested students.* Students learn more when they are actively engaged in discovery and problem-solving. Collaboration sparks an alertness of mind not achieved in passive listening.
- *Collaboration may help students improve literacy and language skills.* Group work offers students many opportunities to use and improve speaking skills. This is particularly important for second-language learners.
- *Collaborative learning often provides greater psychological health for frustrated learners.* It gives students a sense of self-esteem, builds self-identity, and helps them cope with stress and adversity. It links individuals to group success, so that students are supported, encouraged, and held responsible individually and collectively.
- *Collaboration can help prepare students for today's society.* Team approaches to solving problems, combining energies with others, and working to get along are valued skills in the world of work, community, and leisure.
- *Many times, collaborative learning increases respect for diversity.* Students who work together in mixed-ability groups are more likely to develop friendships with a mixture of races and ethnicities. When students cooperate to reach a common goal, they learn to appreciate and respect each other, from those who are physically handicapped to those who are mentally and physically gifted.
- *Collaborative learning can improve teacher effectiveness with all learners.* Through actively engaging students in the learning process, teachers also make important discoveries about their students' learning. As students take some of the teaching responsibilities, the power of the teacher can be multiplied.

SOCIAL NETWORKS AND UNDERAGE USERS

We all know about young adults and their enthusiasm for social networks. But there are a surprising number of children using social media sites like Facebook. In one fourth grade class we visited, nearly half the children were familiar with Facebook, even though there is little there that is intended for children. (On the surface, games and digital socializing were prime attractions.)

Social networking sites realize that they have a problem and try to protect youngsters from predators. Still, there is general agreement that verifying age over the Internet is someplace between difficult and impossible. Social networking sites generally require users to be thirteen or older; but age inflation is common. (Eighteen is the magic number for some activities.)

Students need to realize that unflattering information, images, and comments they post on Google, Facebook, Twitter, LinkedIn, and other social networking sites are hard to erase. Flickr photos and personal references placed on Wikipedia can be edited and moved around by anyone. So, it should be clear that callous oversharing is a threat to privacy and reputation—as well as future personal, school, and job prospects. Of course, it is technically possible to remove unwanted items, but it is difficult and you can never be completely sure that you have gotten everything.

The fact that the Internet can be a vehicle for damaging someone's life online has serious implications for life off-line. Whatever the required age, children and young adults who pretend to be older can bypass safeguards. And there seems to be little anyone can do about it. In one sense, it is a little like other imaginative technologies in that social media can subvert and disrupt traditional arrangements.

The good news is that many of the parents we interviewed said their youngsters carefully monitored and used social networks responsibly. It is important to note that those parents who were involved in their children's online activities had the most positive views of the results.

SOME SUGGESTIONS FOR ARRANGING THE COLLABORATIVE CLASSROOM

In schools across the country, teachers are spending less time in front of the class and more time encouraging students to work together in small groups. Straight rows are giving way to pods of three, four, or five desks. Of course, collaborative learning is more than rearranging desks. It involves changing how students interact with one another and designing lessons so that teamwork is required to complete assigned tasks.

In the collaborative classroom, group learning tasks are based on shared goals and outcomes. Teachers structure lessons so that to complete a project or activity, individuals have to work together to accomplish group goals. At the same time, they help students learn teamwork skills like staying with the group, encouraging participation, elaborating on ideas, and providing critical analysis.

One of the keys to success is building a sense of cooperation in the classroom. Teachers often start by providing the class with a collaborative activity. The second step is to have groups of three or four students work together on an initial exploration of ideas and information. To encourage group interdependence, teachers can use a small-group version of a strategy like K-W-H-L-S.

- What do we *know*?
- What do we *want to learn*?
- *How will I work with others to learn it?*
- What have I *learned*?
- How have I *shared* what I learned from others?

We suggest that teachers give time for individual and group reflection in the last phase of any collaborative learning activity. This way, struggling learners can analyze what they have learned and identify strengths and weaknesses in the group learning process.

Questions like "Tell something that would help us work better next time," and "How did you contribute to the quality of the group work?" also help in this social-processing stage. Teachers might go on to have student groups engage in activities to reshape their knowledge or information by organizing, clarifying, and elaborating on what has been learned. It's often a good idea to ask student groups to present their findings before an interested and critical audience.

Besides encouraging a sense of group purpose, teachers need to help each student feel that he or she can contribute actively and effectively to class activities. The group may sink or swim together, but individuals are still held accountable for understanding the material. In the collaborative classroom, teachers do more than set standards for group work. They use various assessment tools to evaluate group projects, assignments, and teamwork skills.

To get at individual accountability, consider randomly quizzing group members after group work is completed. Whether or not you decide to interrupt the group is one thing, but providing for some form of individual assessment is a basic requirement.

COLLABORATIVE LEARNING IN THE INTEGRATED EDUCATION CLASSROOM

Collaborative learning has been cited as an instructional strategy that can connect a wide range of struggling students to the regular classroom routines (Slavin, 1990). It has become popular because of its potential to motivate and academically engage all students within a social setting.

It is difficult to arrange educational policy around a single policy—especially one that may change with the political winds. We suggest that a variety of approaches and a wide range of educational research be used to determine what works best in certain situations.

Mounting evidence suggests that integrated applications and collaboration can provide positive outcomes for all students (Holt, 2018).

THE ADVANTAGES OF COLLABORATIVE LEARNING

Understanding the important role that collaboration plays in integrated education provides a way to look at the following benefits:

1. Each person brings to the collaborative process experiences that are shared with others.
2. Support is provided for the classroom teacher.
3. Realistic expectations are determined.
4. Classroom teachers are given support for making modifications.
5. Students can be successful when appropriate modifications are made.
6. Teachers become part of a team in dealing with learning and behavior problems. When students with disabilities are included in the regular classroom, it is important to recognize the fact that they are there to learn the subject matter in collaboration with their peers.

APPLYING THE POWER OF COLLABORATIVE LEARNING

By engaging students in the process of making sense of what they are studying, children have more power to explore freely and meaningfully connect to the subject. In a collaborative environment, the teacher assists children in the construction of meaning and acts more like a facilitator and less like a transmitter of knowledge. When questions that connect to student experience are raised collectively, ideas and strengths can be shared in a manner that supports the cooperative search for understanding.

Group learning thrives in an atmosphere of mutual helpfulness where students know what's happening—and why. Part of creating the right environment means having *the teacher* define objectives, talk about the benefits of collaborative learning, and explain expectations and behaviors such as brainstorming, peer teaching, and confidence building.

A supportive team structure leads to greater productivity for all students. To be successful, each child needs to be held responsible for doing a fair share of the group work. At their best, cooperative groups go beyond individual learning to promote an informal style of asking questions, critical thinking, and creating action plans for all students. Critical analysis and creative problem-solving are a natural part of this active learning process.

COLLABORATION AND HUMAN BEHAVIOR

Cooperation is a major factor that differentiates humans from most other species. In addition, small-group, human collaboration sets up the infrastructure that allows for all sorts of transformational changes. From complicated language patterns to complex technology, teamwork makes all the difference in the world. For example, individuals can't build a large airplane alone; it takes thousands of people cooperating to get the job done.

In most other species, only related individuals help each other. With humans, it is more broadly based and the larger social network can develop knowledge and innovations more easily. In fact, many assume that a shift in social behavior was a key factor in making humans unique.

TEAMWORK SKILLS

No matter how you view collaborative student teamwork, there are many common principles at play. Instruction is not viewed as something that isolated students should have done to them; learning is something done best in association with others. The social context matters. And the way different communications are authored or coauthored affects the understanding, reception, and production of information and knowledge.

Teamwork skills do not develop automatically. They must be taught. As group members work together to produce joint work projects, teachers need to quietly help students having problems promote each other's success through sharing, explaining, and encouraging.

Teachers may not be on center stage all the time, but with collaborative learning, they constantly guide, challenge, and encourage students. They can also help build supportive group environments by explaining collaborative procedures to students, monitoring small-group questions, and helping students assess group effectiveness at the end of an activity. To structure lessons so that students work collaboratively with each other requires an understanding of what makes collaboration work.

An important part of collaboration is structuring an environment where group members understand they are connected with each other in a way that one student cannot succeed unless everyone succeeds. (*Your success benefits me and my success benefits you.*) Group goals and tasks must be designed and carefully communicated so that students believe they share a common fate (*We all sink or swim together in this class.*)

When a team is solidly structured, it tells students that each member has a unique contribution to make to the joint effort (*We cannot do it without you*).

Crafting group work that supports learning for all students requires content and activities that supports cohesive small groups and meets the needs of individuals. Some teachers use some form of collaborative learning in pairs or groups of three or four students about half of the time. Others may set aside less time for cooperative group work. However you set it up, collaborative learning can help your students move beyond competitive and individualistic goal structures.

TEACHING SUGGESTIONS FOR USING COLLABORATIVE LEARNING

- Use your existing lessons, content, and curricula and structure them in cooperative groups. Take any lesson in any subject area with a student of any age and structure it collaboratively.
- Tailor collaborative-learning lessons to your unique instructional needs. This may mean that you may need to provide additional time for planning.
- Diagnose the problems some students may have in working together and intervene to increase the effectiveness of the group process.
- Teach students the skills they need to work in groups. Social skills do not magically appear when collaborative lessons are employed. Skills such as "use quiet voices," "stay with your group," "take turns," and "use each other's names" are the beginning collaborative skills.
- It is important to have students discuss how well their group is doing. Groups should describe what worked well and what was harmful in their team efforts.

ARRANGING THE CLASSROOM FOR COLLABORATIVE LEARNING

Effective teachers know that an important step in changing student interaction is changing the seating arrangement. Architecture and the organization of our public and private spaces strongly influence our lives at every level. The same principle applies to schools and individual classrooms.

The way teachers arrange classroom space and furniture has a strong impact on how students learn. When desks are grouped in a small circle or square, or when students sit side-by-side in pairs, collaborative possibilities occur naturally. Straight rows send a very different message.

A classroom designed for student interaction makes just about anything more interesting. The way you design the interior space of your classroom helps focus visual attention. It also sets up acoustical expectations and can help control noise levels. Natural lighting, carpets, comfortable corners, occasional music, and computers that are arranged for face-to-face interaction can all help set the general feelings of well-being, enjoyment, and morale.

Classroom management is actually easier if students know that they can't shout across the classroom but they can speak quietly to one, two, or three others, depending on the size of the small group. Even many questions that students are used to asking the teacher can come after asking one or two peers. All students benefit in this kind of group-learning situation, even the most reluctant student.

As students engage in collaborative learning, they should sit in a face-to-face learning group that is as close together as possible. The more space you can put between groups, the better. From time to time, it is important to remix the groups so that everybody gets the chance to work with a variety of class members.

The physical arrangement should allow you to speak to the whole class without too much student movement. Struggling students benefit from this grouping arrangement. Teachers can give students more of their attention and better differentiate instruction. When the whole class is together, you should be able make eye contact with every student in every group without anyone getting bent out of shape or moving desks.

WAYS TEACHERS CAN ORGANIZE FOR COLLABORATIVE LEARNING

1. Formulate objectives.
2. Decide on the size of groups, arrange the room, and distribute the materials students need.
3. Explain the activity and the collaborative group structure.
4. Describe the behaviors you expect to see during the lesson. These group behaviors may include the following:
   - share ideas
   - respect others
   - ask questions
   - stay in your group
   - give encouragement
   - stay on task

- use quiet voices
5. Assign Roles.

   Classes new to the collaborative approach sometimes assign each member of the group a specific function that will help the group complete the assigned task. For example: the *reader* reads the problem, the *checker* makes sure that it is understood, the *animator* keeps it interesting and on task, and the *recorder* keeps track of the group work and tells the whole class about it. If you have groups of three, then everyone can share the animator's role. Struggling learners need to be included in these roles.

   No matter how you set up collaborative learning, group achievement depends on how well the group does *and* how well individuals within the group learn.
6. Monitor or intervene when needed.

   While you conduct the lesson, check on each learning group when needed to improve the task and teamwork. Bring closure to the lesson.
6. Evaluate the quality of student work.

Ensure that students themselves will evaluate the effectiveness of their learning groups. Have students construct a plan for improvement. Be sure that all students are on task. Groups may be evaluated based on how well members performed as a *group*. The group can also give individuals specific information about their contribution. Groups can keep track of who explains concepts, who encourages participation, who checks for understanding, and who helps organize the work.

PROBLEM SOLVING IN A SOCIAL SETTING

Problem solving and collaboration are common themes that cut across the content standards and the curriculum. But learning to solve problems in school is often different from the way it happens outside of school. When they get out in the real world, students may feel lost because nobody's telling them what to solve. In real life, we are usually not confronted with a clearly stated problem with a simple solution.

Often, we have to work with others to just figure out what the problem is. The same thing is true when it comes to asking and answering questions. When teachers and students can relate to other people, it can bring out the best in themselves and in others.

Knowledge is constructed over time by learners within a meaningful social setting. Students talking and working together on a project or problem experience the fun and the joy of sharing ideas and information. When students construct knowledge together, they have opportunities to compare knowledge, talk it over with peers, ask questions, justify their position, confer, and arrive at a consensus. Even students who usually struggle with a project will feel a sense of belonging to the group.

Collaboration will not occur in a classroom that requires students to always raise their hands to speak. Active listening is not sitting quietly as a teacher or another student drones on. It requires spontaneous and polite interruptions where everyone has an equal chance to speak and interact. Just let others complete a thought and don't break into the conversation in mid-sentence. Try to get everyone to ask a question or make a comment.

It may be best to *not* make students put their hand up first. Encourage the more talkative class members to let everyone make a contribution before they make another point. The inattentive listener may need to assume a leadership role and help monitor the discussion.

Collaborative learning will involve some change in the noise level of the classroom. Sharing and working together even in controlled environments will be louder than an environment where students work silently from textbooks. With experience, teachers learn to keep the noise constructive. Whether you are a parent or a teacher, you know that a little reasoning (regarding rules) won't hurt children. Responsible behavior needs to be developed and encouraged with consistent classroom patterns.

When collaborative problem-solving is over, students need to spend time reflecting on the group work. This is a basic question for the end: "What worked well and how might the process be improved?" Students and teachers need to be involved in evaluating learning products and the collaborative group environment.

Effective interpersonal skills are not just for a collaborative learning activity; they also benefit students in later educational pursuits and when they enter the workforce. Social interactions are fundamental to negotiating meaning and building a personal rendition of knowledge. Mixed-ability learning groups have proven effective across the curriculum. It is important to involve students in establishing rules for active group work.

CLASS RULES FOR COLLABORATIVE LEARNING

Rules should be kept simple and might include the following:

- Everyone is responsible for his or her own work.
- Productive talk is desired.
- Each person is responsible for his or her own behavior.
- Try to learn from others within your small group.

- Everyone must be willing to help anyone who asks.
- Ask the teacher for help if no one in the group can answer the question.

Group roles and individual responsibilities also need to be clearly defined and arranged so that each group member's contribution is unique and essential. If the learning activities require materials, students may be required to take responsibility for assembling and storing them. Avoid getting too many materials too fast. Three or four problems with materials are enough for the struggling learner. All students want to be using materials. Unlike competitive learning situations, the operative pronoun in collaborative learning is "we," not "me."

DESIGNING COLLABORATIVE GROUP LESSONS

During the initial introduction of a lesson, you can help your students understand what it is they're supposed to do by establishing guidelines on how the group work needs to be conducted. Present and review the necessary concepts or skills with the whole class and pose a part of the problem or an example of a problem for the whole class to try. Provide a lot of opportunities for your students to discuss a wide range of issues meaningful to them. Present the actual group problem after you finish the conceptual overview. Then, encourage your students to discuss and clarify the problem task.

When they're ready, students start to work collaboratively to solve problems. You'll need to listen to the ideas of the different teams and offer assistance when you detect that some of them are getting stuck. You're also responsible for designing extension activities just in case the faster teams finish early.

There are different ways of handling teams that are stuck. One way would be for you to help them discover what they know, so far, and then, pose a simple example, or, perhaps, point out a misconception or erroneous idea that may be getting in their way. For example, team members may have trouble getting along with each other or focusing on the one very specific task they're supposed to be doing.

Pull their energies together by asking them simple questions like these: "What are you supposed to be doing now?" "What is your team's task?" "How will you get organized from where you are now?" "What materials do you need?" "Do you think you have enough time to cover everything you set out to do?" "Do you know who will do what?" This is very helpful for disengaged students.

After students complete the problem task and group exploration stages, they will need to meet again as a whole class to summarize and present their findings. Each team needs to present their solutions and tell their classmates how they worked toward their resolution as a group.

You or the other students in class could very well ask questions like these: "How did you organize the task?" "What problems did your team have?" "What method did you use?" "Was your group method effective? Why or why not?" "Did anyone have a different method or strategy for solving the same or similar problem?" "Did your team think that your solution made sense?"

Encourage your students to listen and respond to their classmates' comments. You may, in fact, point out to them that they could earn participation points in this exercise by responding precisely to their classmates' remarks and building upon them. Ask the recorder in the group to make notes on the chalkboard and write down students' responses to help summarize class data at the end of the lesson.

HELP ALL LEARNERS SUCCEED

Flexible grouping strategies allow for high levels of flexibility and creativity that can be used with students of all ability levels. Here, we suggest collaborative strategies for helping all students succeed in a regular classroom setting (U.S. Department of Education, 2009).

1. Assign students to flexible groups.

Organize the class into four-student groups. One way to accomplish this is to use partner groups. Rank the class from "the most prepared" to "the least prepared" for the subject (e.g., number the students from 1 to 30). Next, divide them into subgroups (1 to 10, 11 to 20, 21 to 30) so that the groups are similar to the traditional high-, middle-, and low-ability groups. Finally, achieve mixed grouping by assigning the top student in each of the three groups to one group, the second-highest student in each to another, the third-highest to another, and so on.

You will then have students 1, 11, students 10, 20, and 30. In this way, the mixed groups should comprise students who are sufficiently different in ability that can benefit from each other's help, but not so different that they find one another intimidating.

Inform students of their group assignments, and tell them that they are partners and must help each other as needed, whether by reading each other's work before it is turned in, answering questions regarding assignments, showing a partner how to do something, or discussing a story and sharing their ideas. Let them know that this is only one of many grouping arrangements that you will be using. Grouping procedures may be based on skills, levels, or interests. Collaborative groups can be based on tasks or goal achievement.

2. Focus on the needs of students.

Students learn best when they satisfy their own motives for learning the material. Some of these motivations include the need to learn something in order to complete a particular task or activity, the need for new experiences, and the need to be involved and to interact with other people.

3. Make students active participants in learning.

Students learn by doing, making things, writing, designing, creating, and solving problems. The first step is to honor the different ways that students learn.

4. Help students set achievable goals for themselves.

Often, students fail to meet unrealistic goals. Encourage students who are struggling to focus on their continued improvement. Help students to evaluate their progress by having them critically look at their work and the work of their peers.

5. Work from students' strengths and interests.

To help students find areas where they have a special talent or interest, such as sports, art, or car mechanics, teachers may give them interest inventories. Ultimately, each student selects an area of special interest or curiosity and discusses the topic with the teacher and their peers. Then, they begin a search for more information, which may lead to a group project or a team presentation.

6. Be aware of the problems students are having.

Meet with your students one-on-one for a brief conference. It's helpful to tape the conversation so you have an oral explanation of their understandings. Play the tape for your student and ask questions if the student is confused.

7. Organize a conducive team-meeting environment.

Oftentimes, students are easily distracted by the sights and sounds in the room. Choose an area of the classroom that presents the fewest distractions and keep visual displays purposeful.

8. Incorporate more time and practice for students.

Students who are having difficulties remembering skills need small doses of increased practice throughout the day. This increases performance.

9. Provide clarity.

Clarity is achieved by modeling and using open-ended questions so you can adjust your approach to different students.

10. Intervene early and often.

The key to intervention strategies is identifying students who need extra help and providing ways for struggling students to receive support.

COLLABORATIVE LEARNING ACTIVITIES

A carefully balanced combination of integrated education, subject-matter knowledge, knowledge of the students, instruction, self-monitoring, and active group work helps meet diverse needs of all students. The activities suggested here are designed to provide a collaborative vehicle for active learning in math and science.

Activity 1: Build a Square

This activity can be used with just about any elementary or middle-school class. It relates to the geometry standard in math and the communication standard in both math and science.

Materials

An envelope containing five puzzle pieces. Either the teacher or the students can make the puzzle pieces from three-inch squares of index cards. Cut the index cards into three pieces. Place the puzzle pieces in an envelope.

Procedures

Five people around your table will all make an individual square. Each square has three pieces. Each group leader opens the envelope and passes out the puzzle pieces like a deck of cards so each person has three pieces. *No one is allowed to talk or gesture during this activity.* Group members can pick up a piece and offer it to someone in their group. They can take it or refuse it. *No reaching over and taking pieces!* Raise your hands when all the exchanges have been made and all of the five squares are completed. Students work together silently. They are eagerly trying to get their squares done. It is almost impossible for a struggling student to fail when the whole group is focused.

Evaluation

Next, try the activity again, only this time everyone is allowed to talk. Take time to have a class discussion concerning the problems your group had. Suggest ideas that would make this activity work better.

Activity 2: Back-to-Back Communication

This can be used from third grade on up through middle school. It relates to the geometry and measurement standards in math and the communication standard in both math and science.

Materials

Cut out shapes that can be easily moved on a desk. Make many geometric shapes and make each shape a different color. Colored paper is the simplest material, but Attribute Blocks or Pattern Blocks can also be used.

Procedures

1. Give each group of two people two envelopes with matching sets of shapes. Students at a beginning level should get five or six shapes. More advanced students might use a dozen or more.
2. Have children get into groups of two, seated back-to-back, with their envelopes in front of them.
3. Tell students one of them is the teller and the other listens and tries to follow directions exactly.
4. The teller arranges one shape at a time in a pattern. As the teller does this, he or she gives the listener exact directions—what the shape is and where to place it.
5. When the pattern is finished, have students check how well they have done.
6. Switch roles and do it again.

Evaluation

Have students explain the activity to a partner and describe what was difficult, and how they did it.

Activity 3: Investigate Your Time Line

This works from third grade on up through middle school. Among other things, it relates to the measurement standard in mathematics.

Objectives

1. Each group of four or five students makes a time line of the ages of the people in their groups and the events in their lives.
2. Students will compare the events in their lives with those of other students. (For example: "The most important event for me when I was five years old was . . .")
3. Students record and report the results.

Background Information

- A time line can show different cultural and ethnic patterns.
- Students are able to see how maturity affects decisions.
- A time-line exercise is designed to find out how time changes students' math and science perceptions.

Materials

A thirteen-foot-long piece of butcher paper for each group, rulers, fine-point markers, and a time-line model prepared by the teacher to post on the board for the students to use as a model.

Procedures

1. The teacher will explain that the students will be working in collaborative groups to make time lines of the ages and lives of the people in their groups.
2. The teacher will divide students into groups of four or five students.
3. The teacher and students will pass out the materials to each group.

The teacher will explain his or her model time line and give students directions for making their own time lines:

- Students will find out the ages of the people in their group: who is the oldest, next oldest, youngest, and so on.
- Students will start the time line on January first of the year that the oldest person in the group was born.
- Students will end the time line on the last day of the current year.

- Each student will use a different color marker to mark off each year.
- Each year equals one foot and an inch equals a month.
- At the bottom of each year, the students will write the important events in their lives.
- A color key with the colors of markers and each student's name will identify the student. Students can put a dot or star by the important events in their lives such as birthdays, birth of siblings, and other important events in their lives.

Evaluation

A volunteer from each group will present their group's time line and post it up on the classroom bulletin board.

Activity 4: Bridge Building

This is intended for third through ninth grades. It supports the measurement, geometry, and communication standards in math and the physical science, investigation, and experiment standards in science. Bridge Building is an interdisciplinary math and science activity that reinforces skills related to communication, group process, social studies, language arts, technology, and the arts.

Materials

Lots of newspaper; masking tape; one large, heavy rock; and one cardboard box. Have students bring in stacks of newspaper. You will need approximately a one-foot pile of newspapers per small group.

Procedures

A. For the first part of this activity, divide students into groups of about four. Each group will be responsible for investigating one aspect of bridge building.

Group One: Research
This group is responsible for going to the library and looking up facts about bridges, collecting pictures of all kinds of bridges, and bringing back information to be shared with the class.

Group Two: Aesthetics, Art, Literature
This group must discover songs, books, paintings, artwork, and so forth, which deal with bridges.

Group Three: Measurement, Engineering

This group must discover design techniques, blueprints, angles, and measurements of actual bridge designs. If possible, visit a local bridge to look at the structural design and other features.

Have the group representatives get together to present their findings to the class. Allow time for questions and discussion.

B. The second part of this activity involves actual bridge construction with newspapers and masking tape.

1. At the front of the room, assemble the collected stacks of newspaper, tape, the rock, and the box. Divide the class into groups. Each group is instructed to take a newspaper pile to their group and several rolls of masking tape. Explain that the group will be responsible for building a stand-alone bridge using only the newspapers and tape. The bridge is to be constructed so that it can support the large rock and so that the box can pass underneath.
2. Planning is crucial. Each group is given *ten minutes of planning time* in which they are allowed to talk and plan together. During the planning time, they are not allowed to touch the newspapers and tape, but they are encouraged to pick up the rock and make estimates of how high the box is, make a sketch of the bridge, or assign group roles of responsibility.
3. At the end of the planning time, students are given about *fifteen minutes* to build their bridge. During this time, there is no talking among the group members. They may not handle the rock or the box—only the newspapers and tape. A few more minutes may be necessary to ensure that all groups have a chance of getting their construction to meet at least one of the two "tests" (rock or box). If a group finishes early, its members can add some artistic flourishes to their bridge or watch the building process in other groups. (With children, you may not want to stop the process until each group can pass at least one "test.")

Evaluation

Stop all groups after the allotted time. Survey the bridges with the class and allow each group to try to pass the two tests for their bridge. They get to pick which test goes first. Does the bridge support the rock? Does the box fit underneath? Discuss the design of each bridge and how they compare to the bridges researched earlier. Try taking some pictures of the completed work before you break them down and put them in a recycling bin. Awards could be given for the most creative bridge design; the sturdiest, the tallest, and the widest bridge; the best group collaboration; and so on. Remember, each group is proud of their bridge.

ATTITUDES CHANGE AS CHILDREN COLLABORATE

Some students may require a shift in values and attitudes if a collaborative learning environment is to succeed. The traditional school experience has taught many students that the teacher is there to validate their thinking and direct learning. Getting over years of learned helplessness may take time.

Attitudes change as students learn to work cooperatively. As they share rather than compete for recognition, struggling students find time for reflection and assessment. Small groups can write collective stories, edit each other's writing, solve problems, correct homework, prepare for tests, investigate questions, examine artifacts, work on a computer simulation, brainstorm an invention, create a sculpture, or arrange music. Working together is also a good way for students to synthesize what they have learned, collaboratively present to a small group, coauthor a written summary, or communicate concepts.

It is important that students understand that simply "telling an answer" or "doing someone's work" is not helping a classmate learn. Helping involves learning to ask the right question to help someone grasp the meaning or it involves explaining with an example. These understandings need to be actively and clearly explained, demonstrated, and developed by the teacher.

A major benefit of collaborative inquiry is that students are provided with group stimulation and support. The small group provides safe opportunities for trial and error as well as a safe environment for asking questions or expressing opinions. More students get chances to respond, raise ideas, or ask questions. As each student brings unique strengths and experiences to the group and contributes to the group process, respect for individual differences is enhanced.

The group also acts as a motivator. We all feel a little nudge when we participate in group activities. Many times, ideas are pushed beyond what individuals would attempt or suggest on their own. Group interaction enhances idea development and students can discover their leadership skills when they become teachers as well as learners. In addition, the small-group structure extends children's resources as they are encouraged to pool strategies and share information.

If the group is small enough, it's hard for the more withdrawn students not to participate. Students soon learn that they are capable of validating their own values and ideas. This frees teachers to move about, work with small groups, and interact in a more personal manner with students.

SUMMARY, CONCLUSION, AND LOOKING AHEAD

Collaborative learning is a proven teaching strategy in which small teams of students use a variety of learning activities to gain an understanding of the subject matter. In a collaborative classroom, individuals at various skill levels are encouraged to work together to reach common goals. Along the way, students also learn to take responsibility for themselves and others.

When teachers build on the social nature of learning, students usually become more motivated to explore meaningful inquiry and problem-solving. As students learn to cooperate and work in small, mixed-ability groups, they can also take on more responsibility for themselves and help others to learn. The basic idea is that the social nature of learning builds on the continuous interaction between perception and action.

From the content standards to the Common Core, we find suggestions for supporting and understanding the process of teamwork, creativity, and innovation. Clearly, students' interest in their peers is a resource that can stimulate the learning process through group activities. Whether it is science, the arts, or anything else, the best collaborative lessons make us laugh a little and think a lot, and they help us dream.

New technological tools can help with developing effective groups by encouraging online interpersonal communication that connects to off-line relationships. Also, tech tools can help collaborative groups evaluate data and cut through today's information clutter. Even with the best of intentions, it is wrong to make decisions in an intellectual vacuum or without involving our peers. Together, we can weigh the odds, calculate the risks, and figure out the trade-offs. A good motto: do your best today; do better tomorrow.

For a long time, most subjects have depended on learning the facts before forming opinions. Of course, it has always been fine to put forward a hypothesis to be proven or disproven. But we now live in a "post-truth age" and too often cherry-pick "facts" to support preformed opinions. And although it may be hard to prove something in absolute terms, it is still best to reflect on what we are doing, check in with peers, and make decisions by looking at realistic probabilities.

Imaginative change begins with imaginative ideas. Virtual instruction has its place, but when it comes to helping students generate new ideas, you can't substitute technology for live teachers. In fact, there is no strong research that suggests that online work or courses are as effective as face-to-face learning. Although the group "sinks or swims" together, individuals are held accountable because students receive information and feedback from peers and from their teacher. In a collaborative classroom, many activities are arranged in a way that allows for a merger of academic and social learning experiences.

In collaborative lessons, there is respect for individual group members' abilities and contributions. Cutting corners and costs is one thing; benefiting students, quite another. In school and in life, individuals can usually achieve more by working in collaboration with others.

What can teachers do? One thing is to tell students *what to do*, rather than *how* to do something; this way they can surprise you with their ingenuity. Also you can help students prepare for the ambiguity and the unexpected events that you can be sure are coming along sooner or later. To paraphrase Pasteur, *change favors the prepared mind.*

> The age of cooperation is approaching.... [T]eachers and administrators are discovering an untapped resource for accelerating students' achievement: the students themselves.
>
> —Robert Slavin

REFERENCES

Baker, M., Andriessen, J., and Jarvela, S. (Editors). (2013). *Affective Learning Together: Social and Emotional Dimension of Collaborative Learning.* London: Routledge.

Bishop, P., and Allen-Malley, G. (2004). *The Power of Two: Partner Teams in Action.* Westerville, OH: National Middle School Association.

Bellanca, J., and Stirling, T. (2011). *Classrooms without Borders: Using Internet Projects to Teach Communication and Collaboration.* New York: Teachers College Press.

De Prete, T. (2013). *Teacher Rounds: A Guide to Collaborative Learning in and from Practice.* Thousand Oaks, CA: SAGE.

Dixon-Krauss, L. (1996). *Vygotsky in the Classroom.* New York: Longman.

Hargreaves, A., and O'Connor, M. (2018). *Collaborative Professionalism: When Teaching Together Means Learning for All.* Thousand Oaks, CA: SAGE.

Heyes, C. (2018). *Cognitive Gadgets: The Cultural Evolution of Thinking.* Cambridge, MA: Harvard University Press.

Holt, M. (2018). *Collaborative Learning as Democratic Practice: A History.* Urbana, IL: National Council of Teachers of English (NCTE).

Jacobs, G., Power, M., and Loh, W. I. (2016). *The Teachers' Sourcebook for Cooperative Learning: Practical Techniques, Basic Principles, and Frequently Asked Questions.* New York: Skyhorse Publishing.

Johnson, D. W., Johnson, R. T., and Holubec, E. J. (2009). *Circles of Learning: Cooperation in the Classroom.* 6th edition. Edina, MN: Interactive Book Company.

National Research Council. (1996). *National Science Education Standards.* Washington, DC: National Academies Press.

Sharratt, L., and Planche, B. (2016). *Leading Collaborative Learning: Empowering Excellence.* Thousand Oaks, CA: Corwin Press.

Slavin, R. (1990). *Cooperative Learning: Theory, Research, and Practice.* 2nd edition. Upper Saddle River, NJ: Prentice Hall.

Strebe, J. (2018). *Engaging Students Using Cooperative Learning.* 2nd edition. New York: Routledge.

Stronge, J. (2018). *Qualities of Effective Learners.* 3rd edition. Alexandria, VA: Association for Supervision and Curriculum Development (ASCD).

Topping, K., Buchs, C., Duran D., and van Keer, H. (2017). *Effective Peer Teaching: From Principles to Practical Implementation.* Oxan, UK: Routledge.

U.S. Department of Education (2009). *Evaluation of Evidence-Based Practices in Learning: A Meta-Analysis and Review of Online Learning Studies.* Washington, DC: U.S. Department of Education.

Vygotsky, L. S. (1978). *Mind in Society: The Development of Higher Psychological Processes.* Edited by M. Cole, Vera John-Steiner, S. Scribner, and E. Souberman. Cambridge, MA: Harvard University Press.

Chapter 3

Communications Technologies

> New technologies have been disrupting existing equilibria for centuries, yet balanced solutions have been found before.
>
> —Pamela Samuelson

Available technologies have always been tilting the classroom scene back and forth. Whether it is in or out of the classroom, new ways of communicating and relating to information have frequently required a break from habit. When a new medium emerges, it often generates new ways of thinking and the opening of new realities. A good example can be found in today's online world where the whole world may be watching and connecting names to Facebook pictures.

The arrival of a new medium has always been exhilarating, frightening—and finally, it's just part of life. A few decades into the twentieth century, it was the telephone that began to be taken for granted. By the 1950s it was television. Now, many children and young adults see digital devices and the Internet as a natural way to extend their information and communication reach.

It is important to recognize the fact that each new technology creates a new human environment, which leads to new ways of thinking. Emerging electronic information and communications technology can conjure up new environments for critical thinking, creativity, and teamwork. When used intelligently, they can help you do all kinds of things better.

In a school setting, digital technologies can encourage analytical thinking and help students connect subjects to collaborative inquiry (Sloman and Fernbach, 2017). But still, in some ways, the same technologies and their applications encourage uncritical users to know more and more about less and less.

As educators increasingly take an active role in the development of educational technology, there is more of a reasoned curriculum connection. And the process itself can have something of a liberating effect on the imagination of all involved. Also, new technological possibilities can encourage new habits of the mind and fresh perspectives. Getting all of this right requires everyone involved to view teachers' professional development as a necessity (rather than a luxury).

Unleashing the potential of digital media requires serious thinking, research, and experimentation to make the connections between technology and the characteristics of effective instruction and educational technology. Future directions are open to question. But what is clear is that new technologies herald the arrival of a different educational configuration.

POWERFUL TOOLS AND UNEXPECTED POSSIBILITIES

Done wrong, activities that rely on digital technology can be sad and lonely. When done right, the same technology can help an active meaning-centered curriculum to flourish in (and beyond) the classroom. Either way, the coming together of technologies like computers, video, satellites, and the Internet is both evolutionary and revolutionary.

Where do creative solutions come from and what sparks chains of creation? Computers and their associates are potentially powerful tools for communications, academic work, and innovation. Digital technology has the power to move literacy and learning patterns off established roads. By motivating students through the excitement of discovery, a wide assortment of technological tools can assist the imaginative spirit of inquiry and make lessons sparkle. It can even put students right out on the edge of discovery— where truth throws off its various disguises.

As human horizons shift, flexible drive and intent are required for innovation and progress. Technology can add power to what we do and help us kick against educational boundaries. The vivid images of electronic media can stimulate students as they move quickly through mountains of information, pulling out important concepts, and following topics of interest.

The online process changes students' relationship to information by allowing them to personally shift the relationship of knowledge elements across time and space. Learners can follow a topic between subjects, reading something here, and viewing a video segment there. All of this changes how information is structured and how it is used. It also encourages students to take more responsibility for their own learning.

The negative side of smartphone use and online activities is the practice of multitasking and the shortening of attention spans. This gets in the way of reflection, contemplation, and even thoughtful decision-making. Instead, individuals exist in the realm of constant messages, new contacts, and continuous information flow (Mishra and Henriksen, 2018).

> All media as extensions of ourselves serve to provide new transforming vision and awareness.
>
> —Marshall McLuhan

It is often difficult to detect the subtle happenstance and how we make room in our own lives for positive accidents to happen. Being exposed to different experiences and paying attention to what's going on in the world help by opening up all kinds of serendipitous possibilities. Training the eye to notice things goes a long way toward making unpredictable advances happen. Each new finding can open up fresh questions and possibilities—breaking the habits that get in the way of creative thinking and change. Prescription for thinking in the future: following curiosity, leaving doors open, using technological tools, and making room for good luck to happen.

THE TECHNOLOGICAL DIMENSIONS OF LEARNING IN A NEW ERA

At school, the challenge is getting students to apply the same level of intensity to their schoolwork that they apply to social media and video games. Building a dynamic model of learning requires making good use of everything available. But action without vision can be a nightmare, and vision without action often leads nowhere. One thing is certain—new technology is bound to generate new ways of thinking, learning, and working.

The playful gleam in the eye is often an engine of progress. There are multiple tools and modes of expression that schools need to build on to promote the multitude of strengths and imaginations found in all students. Many schools are mired in unproductive routines that prevent teachers from making creative breakthroughs. Still, we should recognize that it's difficult to redesign the plane when you're the pilot and flying at 30,000 feet.

Educators often don't have enough time to go back to the drawing board and use good data and experience to get it right. It's more than learning how to navigate massive amounts of information; it's how we personally reframe *truth, beauty, and goodness* (Gardner, 2011). Generating new ideas, reflection, and changing approaches takes space, support, and time for collegial professional development to make all the changes required.

Powerful forms of face-to-face learning within schools must not be neglected as we sort out the new media possibilities. Intelligent use of electronic forms of learning have proven to be helpful in improving student learning. But when it comes to the professional development of teachers, their value is not as clear. Electronic learning can, however, make a contribution. And it is a useful supplement to the professional development toolkit.

It is difficult to unravel issues of creativity or analysis without taking into account influential media, like computers and television. They have a tremendous impact on children. We shape them and they shape us. Some are often written about as Lady Caroline Lamb wrote about Lord Byron: *"Mad, bad and dangerous to know"* (Hayle, 2016). Others see technology, such as computers and the Internet, as particularly dangerous enemies, creating a culture of electronic Peeping Toms without a moral foundation.

To be valuable, educational technology must contribute to the improvement of education. Digital devices and their accessories should be designed to help open doors to reality and provide a setting for reflection. By making important points that might otherwise go unnoticed, these technological tools can help students refine and use knowledge more effectively. For example, computers can use mathematical rules to simulate and synthesize lifelike behavior of cells growing and dividing. It's a very convincing way to bring the learning process to life.

Schools can teach students to recognize how technology can undermine social values, human goals, and national intention. They can also help students learn to harness these powerful tools so that they might strengthen and support the best in human endeavors.

It is our belief that when the pedagogical piece is in place, technology can support and strengthen the best in student learning. This can not only change *what* teachers teach but also *how* they go about doing it. The whole process has as much to do with the arts as it does with the sciences.

> The schools must fully acknowledge and embrace their responsibilities to service, values, and the liberal arts—as well as to *reason* and *discovery*.
>
> —paraphrasing Drew Gilpin Faust

THE FUTURE OF NARRATIVE IN CYBERSPACE

The idea of print as an immutable canon may or may not be a historical illusion. One thing for sure, the way print is being mixed electronically with other media changes things. Although the American book industry is rushing into the emerging electronic literary market, book pages made of paper won't go away. And print is here to stay. Even the doomsayers usually use books to put forward their argument that the medium is a doomed and outdated technology. In the future, will books be confined to dusty museum libraries? No, they remain an elegant, user-friendly medium. With books you don't need batteries and don't have to worry about the technological platform becoming obsolete and unusable.

The major goal of any form of reading instruction will continue to be getting students to comprehend what they read.

There are at least two fairly new digital approaches to books that are finding a niche in the literacy universe. One is much like an electronic version of printed books. The other approach to electronic books is interactive and visually intensive. It takes the narrative and places it in randomly accessible blocks of text, graphics, and moving video. With some e-book stories, students must learn to go beyond merely following the action of the plot to learn about characters, explore different ideas, and enter other minds.

An interactive e-book story places students in charge of how things develop and how they turn out. Participants are able to change the sequence or make up a new beginning to a multidimensional story. They can slow up to find out additional information and they can change the ending. Navigating interactive stories with no fixed center, beginning, or end can be very disconcerting to the uninitiated. It requires different "reading" skills.

To make sense of the anarchy and chaos, a reader has to become a creator. This means following links around so that they can discover different themes, concepts, and outcomes. "Interactive Storytime" is an early example. It tells stories with narration, print, music, sound effects, and graphics. Children can click on any object and connect spelling to the pronunciation.

Literature has traditionally had a linear progression worked out in advance by the author. The reader brought background knowledge and a unique interpretation to what the author had written. But it was the author who provided the basic sequential structure that pushed all readers in the same direction. Computer-based, multidimensional literature is quite different. The reader shapes the story line by choosing the next expository sequence from a number of possibilities.

With early versions of interactive, computer-based literature, readers are connected to a vast web of printed text, sound, graphics, and lifelike videos. When key words or images are highlighted on a computer screen, the reader clicks what they want next with a mouse (or use their finger on a touch screen) and the reader hops into a new place in the story, causing different outcomes. With a virtual-reality format, the reader uses their whole body to interact with the story. Whatever the configuration, interactive literature causes the user to break down some of the walls that usually separate the reader from what's being read.

The forking paths in this electronic literature poses new problems for readers—like how do you know when you have finished reading when you can just keep going all over the place. Judy Malloy's *Its Name Was Penelope*, for example, shuffles 400 pages of a fictional woman's memories so that they come together very differently every time you read it. The ambiguity of these programs isn't as bothersome as it used to be.

ONLINE PEER TUTORING

Whether it is *off*-line or *on*line, peer tutoring involves several methods. In *cross-age tutoring,* older students tutor younger students. In *cross-ability tutoring,* a student who has a good grasp of the subject can tutor a student who is struggling. *Reciprocal tutoring* involves a structure for interaction that is tied to specific academic goals.

CONVEYING MEANING WITH POWERFUL VISUAL MODELS

One way to enhance the power and permanency of what we learn is to use visually based mental models in conjunction with the printed word. Inferences drawn from visually intensive media can lead to more profound thinking. In fact, children often rely on their perceptual (visual) learning even if their conceptual knowledge contradicts it. In other words, even when what's being presented runs contrary to verbal explanations, potent visual experiences can push viewers to accept what is presented.

Children can become adept at extracting meaning from the conventions of video, film, or animation—zooms, pans, tilts, fade-outs, and flashbacks. But distinguishing fact from fiction is more difficult.

The ability to understand what's being presented visually is becoming ever more central to learning and to our society. Most of the time, children construct meaning for television, film, computer, or Internet content without even thinking about it. They may not be critical consumers, but they attend to stimuli and extract meaning from subtle messages.

The underlying message children often get from the mass media is that viewers should consume as much as possible while changing as little as possible. How well content is understood varies according to similarities between the viewers and the content. Learners' needs, interests, and age are other important factors to consider.

Sorting through the themes of mental conservatism and material addition requires carefully developed thinking skills.

Meaning in any medium is constructed by each participant at several levels. For better or for worse, broadcast television used to provide us with a common culture. When viewers share a common visual culture, they must also share a similar set of tools and processes for interpreting these signals (construction of meaning, information processing, interpretation, and evaluation).

The greater the experiential background in the culture being represented, the greater the understanding. The ability to make subtle judgments about what is going on in any medium is a developmental outcome that proceeds from stage to stage with an accumulation of experience.

Relying upon a host of cognitive inputs, individuals select and interpret the raw data of experience to produce a personal understanding of reality. What is understood while viewing depends on the interplay of images and social conditions.

Physical stimuli, human psychology, and information-processing schemes taught by the culture help each person make sense of the world. In this respect, reacting to the content of an electronic medium is no different from any other experience in life. It is just as possible to internalize ideas from electronic visual imagery as it is from conversation, print, or personal experience. It's just that comprehension occurs differently.

Reflective thought and imaginative, active play are important parts of the growth process of a child. Even with a "lean back," passive medium like television, children must do active work as they watch, make sense of its contents, and utilize its messages. With a "lean forward" medium like the Internet, this work is fairly evident. Evaluative activities include judging and assigning worth, assessing what is admired, and deciding what positive and negative impressions should be assigned to the content. In this sense, children are active participants in determining meaning in any medium.

ADULTS INFLUENCE HOW CHILDREN LEARN TO ASSESS MEDIA MESSAGES

Although children learn best if they take an active role in their own learning, parents, teachers, and other adults are major influencers. They can significantly affect what information children gather from television, film, or the Internet. Whatever their age, critical users of media should be able to

- understand the grammar and syntax of a medium as expressed in different program forms.
- analyze the pervasive appeals of advertising.

- compare similar presentations with similar purposes expressed in different media.
- identify values in language, characterization, conflict resolution, and sound/visual images.
- utilize strategies for the management of duration of viewing and program choices.
- identify elements in dramatic presentations associated with the concepts of plot, story line, theme, characterizations, motivation, program formats, and production values.

Parents and teachers can affect children's interest in media messages and help them learn how to process information. Good modeling behavior, explaining content, and showing how the content relates to student interests are just a few examples of how adults can provide positive viewing motivation. Adults can also exhibit an informed response by pointing out misleading messages—without building curiosity for undesirable programs.

The viewing, computer, and Internet using habits of families play a large role in determining how children approach a medium. The length of time parents spend watching television, the kinds of programs viewed, and the reactions of parents and siblings toward programming messages all have a large influence on the child. If adults read and there are books, magazines, and newspapers around the house, children will pay more attention to print. Influencing what children view on television or the Internet may be done with rules about what may or may not be watched, interactions with children during viewing, and the modeling of certain content choices. Families can also seek a more open and equal approach to choosing television shows—interacting before, during, and after the program.

When it comes to the Internet, keep the computer in the family room rather than in a private isolated space. Time limits must be placed on the use of today's electronic gadgets. Parents can organize formal or informal group activities outside the house that provide alternatives to Internet use or TV viewing.

Whether co-viewing or not, the viewing choices of adults (parents, teachers, etc.) in a child's life set an example for children. If, for example, parents are heavy watchers of public television or news programming, then, children are more likely to respond favorably to this content. If parents make informed, intelligent use of the Internet, then, children are likely to build on that model. Influencing the settings in which children watch TV or use the computer is also a factor. Turning the TV set off during meals, for example, sets a family priority.

COMMUNICATIONS TECHNOLOGIES AND PUBLIC CONVERSATIONS

A democratic community is defined by the quality of its educational institutions and its public conversations. Democracy often becomes what it pays attention to. American national values, supported by our constitution, require an educated citizenry that can think, respond to leaders, and are willing to actively go beyond the obvious. Patriotism isn't just the flag and stern rhetoric; it's a thinking, decent, and literate society. Exercising citizenship in a world of accelerated change requires the preservation of our human values.

Ignoring the societal implications of technology means ignoring looming changes. Whether it's technologically induced passivity or the seductive charms of believing in simplistic technological solutions, it is only through the educational process that people can gain a heightened awareness of bright human and technological possibilities. The question that we need to answer is this: How might the technology be used to spark a renaissance in human learning and communication?

The long-term implications of recent changes in information and communications technologies are important, if not frightening. The convergence of technologies is causing a major change in societal behaviors, lifestyles, and thinking patterns. With few people monitoring digital technology or theorizing about its effects on health, the human race is being forced to swim in an electronic sea of information and ideas. In today's world, there is little question that reality is being shaped by electronic information and electronic illusions.

ACTIVITIES FOR MAKING SENSE OF VISUAL MEDIA

1. Help students critically view what they see.

Decoding visual stimuli and learning from visual images require practice. Seeing an image does not automatically ensure learning from it. Students must be guided in decoding and looking critically at what they view. One technique is to have students "read" the image on various levels. Students identify individual elements and classify them into various categories, and then, relate the whole to their own experiences, drawing inferences, and creating new conceptualizations from what they have learned.

Encourage students to look at the plot and story line. Identify the message of the program. What symbols (or camera techniques, motion sequences, setting, lighting, etc.) does the program use to make its message? What does the director do to arouse interest by using color, sound, and images to distort reality and influence consumers?

We often arrange students in small groups and have them come up with sixty-second commercials to sell an imagined product. The next step is have each group practice and then, act out a sixty-second commercial in front of the class. Discuss the results. Would you buy their product?

Analyze and discuss commercials in different media. How many minutes of ads appear in an hour? How many ads do you have to sort through before you can watch a program or use a search engine to get to some websites? What should be done about the ad glut?

2. Create a scrapbook of media clippings.

Have students keep a scrapbook of newspaper and magazine clippings on computers, the Internet, Facebook, Google, television, and some of the other inhabitants of cyberspace. Newspapers and magazines are a good source of articles; the *New York Times Science* section is a good example for upper-grade students. Ask students to paraphrase, draw a picture, or map out a personal interpretation of an interesting technology article. Share these works with other students.

3. Create new images from the old.

Have students take rather mundane photographs and multiply the images or combine them with others in a way that makes them interesting. Through the act of observing, it is possible to build a common body of experiences, humor, feeling, and originality. And through collaborative efforts, students can expand on ideas and make the group process come alive.

4. Role play communicating with extraterrestrial life.

Directions for students: If extraterrestrial life has already contacted us, think about how to respond. Think creatively and scientifically about how you would explain this planet to some inquisitive aliens. Explain human life and media devices; try to view our planet as a whole.

5. Use debate for critical thought.

Debating is a communications model that can serve as a lively facilitator for concept building. Take a current and relevant topic, and formally debate it online or face-to-face. This can serve as an important speech and language extension. For example, the class can discuss how mass media can support everything from commercialism to public conformity and the technological control of society. The discussion can serve as a blend of technology, social studies, science, and the humanities.

DIFFERENT MEDIA SYMBOL SYSTEMS

Print and visually intensive media take different approaches to communicating meaning. Print relies upon the reader's ability to interpret abstract symbols. A video or computer screen is more direct. Whatever medium is used, thinking and learning are based on internal symbolic representations and the mental interpretation of those symbols. When they are used in combination, one medium can amplify another.

We live in a complex society dependent on rapid communication and information access. Lifelike, visual symbol systems are comprised, in part, of story structure, pace, sound track, color, and conceptual difficulty. Computers, the Internet, television, and digital devices are rapidly becoming our dominant cultural tools for selecting, gathering, storing, and conveying knowledge in representational forms (Levy, 2011).

Various electronic symbol systems play a central role in modern communications. It is important that students begin to develop the skills necessary for interpreting and processing the full range of media messages. Symbolically different presentations of media vary as to the mental skills of processing they require. Each individual learns to use a medium's symbolic forms for purposes of internal representation.

Unlike direct experience, print or visual representation is always coded within a symbol system. Learning to understand that system cultivates the mental skills necessary for gathering and assimilating internal representations. Each communications and information medium makes use of its own distinctive technology for gathering, encoding, sorting, and conveying its contents associated with different situations. The technological mode of a medium affects the interaction with its users—just as the method for transmitting content affects the knowledge acquired.

The closer the match between the way information is presented and the way it can be mentally represented, the easier it is to learn. Better communication means easier processing and more transfer. At its best, a medium gets out of your way and lets you get directly at the issues.

UNDERSTANDING AND CREATING ELECTRONIC MESSAGES

Understanding media conventions helps cultivate mental tools of thought. In any medium, this allows the viewer new ways of handling and exploring the world. The ability to interpret the action and messages requires going beyond the surface to understanding the deep structure of the medium. Understanding the practical and philosophical nuances of a medium moves its consumers in the direction of mastery.

Simply seeing an image does not have much to do with learning from it. The levels of knowledge and skill that children bring with them to the viewing situation determine the areas of knowledge and skill development acquired. Just as with reading print, decoding visually intensive stimuli and learning from visual images require practice.

Students can be guided in decoding and looking critically at what they view. One technique is to have students "read" the image on various levels. Students identify individual elements, classify them into various categories, and then, relate the whole to their own experiences. They can then draw inferences and create new conceptualizations from what they have learned.

Planning, visualizing, and developing a production allow students to critically sort out and use media techniques to relay meaning. Young producers should be encouraged to open their eyes to the world and visually experience what's out there. By realizing their ideas through media production, students learn to redefine space and time as they use media attributes such as structure, sound, color, pacing, and imaging.

The field of technology and its educational associates are in a period of introspection, self-doubt, and great expectations. In a world awash with different types of educational media, theoretical guidelines are needed as much as specific instructional methods.

It is dangerous to function in a theoretical or research-deprived vacuum because rituals can spring up that are worse than those drained away. As schools are faced with aggressive marketing for electronic devices, we must make sure that a pedagogical plan that incorporates technology is in place first.

For technological tools to reach their promise, close connections among educational research, theory, and classroom practice are required. Across the curriculum, new standards place a high premium on creative and critical thinking. So, it is clear that the curriculum and professional development will pay even closer attention to such skills.

Reaching students requires opening students' eyes to things they might not have thought of on their own. This means using technological tools as capable collaborators for tapping into real experiences, fantasies, and personal visions. This way previously obscure concepts can become comprehensible and deeper at an earlier age.

Technology and metacognitive strategies can come together as students search for data, solve problems, and graphically simulate their way through multiple levels of abstraction. The combination of thoughtful strategies and the enabling features of media tools can achieve more lasting cognitive change and improved performance.

SOCIAL EQUITY AND A MODERN PHILOSOPHY OF TEACHING

When it comes to solving our educational problems, technology isn't the only thing. But it is an important thing. As we put together the technological components that provide access to a truly individualized set of learning experiences, it is important to develop a modern philosophy of teaching, learning, and social equity. While new information and communications technologies have the potential to make society more equal, it sometimes has the opposite effect. As we enter new technological worlds, everyone must have access.

Many American school districts lack the money to train teachers to use digital technology effectively. So, in some schools, children are still taught to use the computer primarily for typing, drill, and workbook-like practice. In contrast, affluent schools often encourage students to use digital technology for creative exploration—like designing multimedia presentations and collaborating with classmates in problem-solving experimentation.

Everyone deserves access to a provocative and challenging curriculum. To become an equal instrument of educational reform, technology must do more than reinforce a two-tier system of education. Otherwise, many children will face a discouraging picture of technological inequality.

There are serious social consequences surrounding the inequalities of access to telecommunications and advanced information technology. Future connections to faculty, other students, databases, and library resources will change the way that information is created, accessed, and transmitted. Those denied access may have their academic ambitions stifled, causing them to fall further and further behind. The challenge is to make sure that this information is available for all in new versions of public libraries.

Digital technology gives us the ability to change the tone and priorities of gathering information and learning in a democratic society. Taking the right path requires learning to use what's available today and building a social and educational infrastructure that can travel the knowledge highways of the future. But whether it's today or tomorrow, innovation is more about small productive teams than individuals working alone.

Working with students online, we have found computer simulations that simulate invisible things like molecular reactions and static electricity. In addition, with role-playing, interactive activities from around the world, it is possible to foster thinking skills and collaboration.

Electronically connecting the human mind to people and global information resources may shift human consciousness in ways similar to what occurred in moving from an oral to a written culture. The ultimate consequences are unclear. But the development of basic skills, habits of the mind, wisdom, and traits of character will be increasingly affected by technology.

The curriculum materials that are most effective and most popular are those that provide for social interaction and problem-solving. Information can be embedded in visual narratives to create contexts that give meaning to dry facts.

Interactive digital technology can challenge students on many levels and even serve as a training ground for responsibility, persistence, and collaborative inquiry. It's relatively easy to buy the hardware and get children interested. The difficult thing is to connect to deep learning in a manner that advances curriculum goals.

LEARNING TO EVALUATE ONLINE ACTIVITIES AND SOFTWARE

Curriculum is a cooperative and interactive venture between students and teachers. On both the individual and social levels, those directly involved with the process must be taken seriously. Working together, they can decide what benefits are gained from particular software programs. After all, the software user is in the best position to decide if the program is taking people out of the process—or whether learners are in control of the computers. Good software programs let students learn together at their own pace—visualizing, talking together, and explaining abstract concepts so that they can relate them to real-life situations.

There are time-consuming evaluation issues surrounding the multitude of software programs to be dealt with. Multimedia, simulation, micro-worlds, word-processing, interactive-literature, spreadsheets, database managers, or expert (AI) systems all increase the potential for influencing impressionable minds. That is too large a universe for the teacher to figure out alone. Teachers can help each other. Another thing that helps the task become more approachable is getting the students to take on some of the responsibility.

Children can learn to critique everything from computer software to Internet websites much as they learned to critique the dominant media of yesteryear—books. One of the first tenets of book review criticism is to critique what is usually taken for granted. Also, having at least a little affection for what is being reviewed helps. Students can do book reports, review computer software, and discuss the quality of sites on the Internet. It helps if students sift through some of the software reviews in newspapers, magazines, and journals—online and off-line.

Examining online activities and software, students can quickly check to see if the flow of a program makes sense. Next, they can try out the software as they think a successful or unsuccessful student might. What happens when mistakes are made? How are the graphics? Do you think you can learn anything from this? Is it exciting?

Without question, teachers and students are the ones who experience the consequences of making good or bad choices in software selection. And they are the ones who most quickly learn the consequences of poor choices.

CRITERIA FOR EVALUATING E-ACTIVITIES

The teacher can apply the same assessment techniques used on other instructional materials when students evaluate courseware. The following list is an example:

1. Does the activity meet the age and attention demands of your students?
2. Does it hold the students' interest?
3. Do the programs, activities, or simulation games develop, supplement, or enhance curricular skills?
4. Does the work require adult supervision or instruction?
5. Children need to actively control what the program does. To what extent does the program allow this?
6. Can the courseware be modified to meet individual learning requirements?
7. Can it be adjusted to the learning styles of the users?
8. Does the program have animated graphics that enliven the lesson?
9. Does it meet instructional objectives and is it educationally sound?
10. Does the activity or program involve higher-level thinking and problem-solving?

STUDENT EVALUATION CHECKLIST

Name of student evaluator(s) _____
Name of software program _____
Publisher _____ Subject _____
1. How long did the activity take? _____ minutes

2. Did you need to ask for help doing the work? _____ yes _____ no
3. What skills do you think the activity tried to teach?

Please circle the word that best answers the questions below:
4. Was it fun to use?   yes   no   somewhat   not very
5. Were the directions clear?   yes   no   somewhat   not very
6. Was it easy to use?   yes   no   somewhat   not very
7. Were the graphics (pictures) good?   yes   no   somewhat   not very
8. Did the activity get you really involved?   yes   no somewhat   not very
9. Did the software allow you to make choices?   yes   no
10. What mistakes did you make? _____
11. What happened when you made a mistake? _____
12. What was the most interesting part of the activity? _____
_____
13. What did you like least? _____
14. What's a good tip to give a friend who's getting started with this activity?
_____

SERVING PEDAGOGY

The curriculum should drive the technology rather than the other way around. The most effective use of such tools occurs when desired learning outcomes are figured out first. Once standards have been set, you can decide what technological applications will help you reach your goals. Technology can help when it serves clear educational goals. But there is much work to be done to put an *e* and a dash in front of education. However, by working together, teachers can learn from each other, push on to other innovations, and build a medium worthy of our students.

Converging communication technologies can serve as a great public resource. Commercial profit alone should not determine how these new technologies will be exploited. The airwaves and the information highways are owned by the public. Emerging public utilities must show a concern for learning and responsible social action. Any new media system is a public trust that should enable students to become intelligent and informed citizens.

In today's world, children grow up interacting with electronic media as much as they do interacting with print or with people. Unfortunately, much of the programming is not only violent, repetitious, and mindless, but also distracts students from more important literacy and physical exercise activities. Dealing with these new digital realities requires a new approach to curriculum and instruction. It also calls for a heightened sense of social responsibility on the part of those in control of programming.

SUMMARY, CONCLUSION, AND LOOKING AHEAD

Learning to separate noise from knowledge is increasingly important in this age of too much information. Making better use of "knowledge machines" (digital technology) means going well beyond electronic workbooks or surfing the Internet to collaboratively explore problems and tinker with disparate ideas. The good news is that it's at least *possible* for technological tools to provide a vehicle for building on students' natural curiosity and promoting real learning through active engagement and collaborative inquiry.

The human mind can create beyond either what it intends or what it can foresee. Combinations of electronic media will be with us no matter how far the electronic environment expands. The problem is being sure that they work to human advantage. The informing power of technological tools can help change schools. It can also make learning come alive and breathe wisdom into instructional activities.

Technological applications go a long way toward explaining the shape of everyone's world today (McAfee and Brynjolfsson, 2017). Along the path to the future, many people have grown used to having technology between them and reality. We shape digital technologies and, at the same time, they shape us. Along the way, many strands of possibilities shape the future of people's lives. Clearly, the focus of education must be shifted in a way that helps prepare students for a future that is vastly different from what their teachers have known.

We know that technologies like the Internet are fragmenting the thought processes of young people. Also, there is less and less time for reflection and quiet thought. But no matter where all the changes take us, reasoning, communication, and teamwork skills are bound to go hand in hand with taking an active part in shaping the future.

Harmonizing students' present with the future requires more than reinventing schools. In America, for example, not only are some schools failing, but the home environment and society are also failing.

Many children are behind on their first day of school, so efforts to improve schooling must extend before and beyond the classroom door. For all learners to thrive academically, it helps to have the benefits of preschool, high-quality teachers, health care, and engaged family support.

Is access to an infinite amount of information infinitely valuable or does it simply lead to the collapse of clear and well-thought-out meaning? Whatever the combination of answers, uncontrolled technological change is a problem—and for new media to contribute to the solution, it has to have a human face. Although the technology-intensive educational path to the future may be bumpy, it doesn't have to be gloomy. Clearly, dedicated teachers who understand and know how to use their high-tech tools can make a real difference.

Although we live in an information age, a clear understanding of the nature of information remains elusive. With or without an understanding of the concept, increasing the information flow presents at least as many problems as it does opportunities (Sax, 2016). About the same thing might be said about technology in general. What's needed is a controlling vision that can help explain concepts related to information, technological innovation, and implications for education.

> We have trod the face of the Moon, touched the nethermost pit of the sea, and can link minds across vast distances. But for all that, it's not so much our technology, but what we believe, that will determine our fate.
>
> —Tim Flannery

REFERENCES

deVries, M. (2018). *Handbook of Technology Education*. New York: Springer.

Foer, F. (2017). *World without Mind: The Existential Threat of Big Tech*. New York: Penguin Press.

Gardner, H. (2011). *Truth, Beauty, and Goodness Reframed: Educating for the Virtues of the Twenty-First Century*. New York: Basic Books.

Hayle, J. (2016). *Lord Byron and Lady Caroline Lamb: Mad, Bad and Dangerous to Know: The Passionate and Public Affair That Scandalised Regency England*. CreateSpace Independent Publishing Platform.

Levy, S. (2011). *In the Plex: How Google Thinks, Works, and Shapes Our Lives*. New York: Simon & Schuster.

Malloy, Judy. (1993). *Its Name Was Penelope*. Cambridge, MA: Eastgate Systems Inc.

McAfee, A., and Brynjolfsson, E. (2017). *Machine, Platform, Crowd: Harnessing Our Digital Future*. New York: W. W. Norton.

McLuhan, M., and Lapham, L. H. (1994). *Understanding Media: The Extensions of Man*. Cambridge, MA: MIT Press.

Mishra, P., and Henriksen, D. (2018). *Creativity, Technology and Education: Exploring Their Convergence*. New York: Association for Educational Communication and Springer.

Samuelson, P. (2018). "Legally Speaking: Copyright Blocks a News-Monitoring Technology." *Communications of the ACM* 61 (7) (July): 24–26.

Sax, D. (2016). *The Revenge of the Analog: Real Things and Why They Matter*. New York: Little, Brown.

Sloman, S., and Fernbach, P. (2017). *The Knowledge Illusion: Why We Never Think Alone*. New York: Riverhead Books.

Chapter 4

Science and Mathematics

Twenty-First-Century Practices in the Classroom

> The important thing about science and mathematics is not so much to obtain new facts as to discover new ways of thinking about them.
>
> —Nobel laureate William Lawrence Bragg

Once someone becomes familiar with some of the concepts and thinking processes associated with science and mathematics, many doors can be opened.

Connecting intuitive ideas about cause-and-effect to the objective realities of science and mathematics can be difficult. So, it is little wonder that so many people are "scienceblind" (Shtulman, 2017). The good news is that observations, inferences, and potential actions can be built on hard facts, inferences, *and* personal opinions.

The scientific method and mathematical problem-solving are useful ways of thinking that can help anyone do a better job of approaching the wide range of possibilities that occur in daily life. Also, when it comes to understanding an increasingly complex and changing world, we can all make better use of reasoning skills based on a range of experience and knowledge.

In fact, the scientific method and mathematical problem-solving are useful ways of thinking that can help anyone do a better job of approaching the wide range of possibilities that occur in daily life. There are many ways to think about "twenty-first-century skills" as there are ways to think about science and mathematics instruction. Here, we limit ourselves to basic content areas, emerging topics, thinking skills, and life skills like adaptability, teamwork, and social responsibility. Close attention is also paid to K–8 classrooms and the negative consequences that stem from separating science and math from humanism.

At all levels, the offshoots of science and math—like information, communications, and technology—are moving closer to the center of the instructional stage. As far as the subjects of science and math are concerned, it is important that students understand at least some of the ideas and applications of experts in the field. It also helps if learners are encouraged to do creative things with their own informed opinions, insights, and conclusions.

Pushing the sphere of what's known is the essence of creativity and innovation. Tomorrow's schools and workplaces highly value those who continually search for fresh ideas and attempt to do new things. Creating fresh ideas and products requires individuals and groups who can go beyond conventional wisdom, take risks, and learn from mistakes (Dweck, 2006).

Innovative change is not always immediate and total. It is possible, for example, to work on the incremental improvements right now, while simultaneously thinking far ahead. Innovation takes many forms. It ranges from Facebook's "move fast and break things" to Bell Labs' "move deliberately and build things." Fast or slow, short- or long-term, scientific reasoning and mathematical problem-solving are keys to technological and engineering success.

Even in the primary grades, it is important for students to realize that scientific and mathematical ideas have a thought-provoking role to play in everything from current affairs to music, sports, and literature (Smith and Stein, 2018).

When it comes to using the imagination to solve problems in new ways, mastering traditional science and math content at school only goes so far. Interacting imaginatively with others and communicating individual values are all part of the process.

TOO IMPORTANT TO BE LEFT TO EXPERTS

The initial mystery that attends any journey is: how did the traveler reach his starting point in the first place?

—Louise Bogan

Many teachers of science and mathematics at the elementary school level have to teach dozens of topics and subjects. So, it should not come as a surprise to find out that many have not received specialist training in every subject. (By the middle grades, specialists are more common.) Still, when it comes to teaching science and math, it is important for all teachers to have the intellectual tools needed to help children learn and apply age-appropriate concepts.

Students at any age can learn to ask big questions and think about grand concepts. In fact, those who work deeply with science and math at an early age are more likely to be creative with these subjects later on (Neagoy, 2012). Technologies like the Internet have made it possible for anyone to learn throughout their lives.

Whether it is individually or in groups, formally or informally, science/math and their technological associates are simply too important to be left to the experts. Only a small percentage of children may become scientists or mathematicians, but everyone must know enough to apply the intellectual tools that are part of these subjects.

Major goals of science and math education in the twenty-first century include helping students develop more self-understanding and move in the direction of responsible citizenship. Along the way, it is important to recognize that new technologies are altering the way knowledge is conveyed. And this does not mean sitting alone in front of a computer screen.

CORE CURRICULUM STANDARDS FOR SCIENCE AND MATHEMATICS

The National Academies Press has published *A Framework for K–12 Science Education: Practices Crosscutting Concepts and Core Ideas* (2012). This publication takes one of the most important steps forward in science and math education since the *National Science Education Standards* (1996) were submitted by the National Committee on Science Education Standards and Assessment and the *Principles and Standards 2001* were released by the National Council of Teachers of Mathematics (Lindquist and Clements, 2001). The NRC constructed the science framework and is working on the next generation of science standards.

The standards bring an imaginative perspective to learning science and math. They emphasize the importance of challenging students in *doing* science and mathematics, not just learning content.

The science framework identifies key scientific ideas and practices that all students should learn. Among other things, it is designed to help students gradually develop, over multiple years, their knowledge of core ideas in four interdisciplinary areas, rather than cultivating shallow knowledge of a wide range of topics.

The latest science framework strongly emphasizes the ways students carry out science investigations and encourages arguments based on evidence (National Academies Press, 2011).

The goal of the new standards is to ensure that all students have an appreciation of the beauty and wonder of science and mathematics. Students should have the capacity to discuss and think critically about science and math-related issues and pursue careers in science, technology, engineering, and math.

Currently, science and math in the United States lack a common vision of what students should know and be able to do. Although legislators may be dazzled by lobbyists for high-tech companies, there is little proof that technology alone improves learning. Also, there are times that, in order to benefit from new technologies, we need to use them less.

SCIENCE, MATH, AND THE SCHOOL CURRICULUM

The evolving nature of science and mathematics is one of the reasons for so much debate within the scientific community and in the general public. As far as classroom-based discussion, experiments, and computations are concerned, students must learn how to collaboratively put the skills they learn into practice. All of these points connect to the subject-matter standards.

Science education is an example: the four science interdisciplinary areas include life science; physical science; earth/space science; and engineering, technology, and the application of science. Some core ideas that cut across these fields include the following: matter, interactions, and energy. Students understand the same is relevant in many fields. These concepts should become familiar as students progress from kindergarten through to twelfth grade.

Both the science and the math standards emphasize key practices that students should learn: asking questions, defining problems, and analyzing and interpreting data. Other important practices include explaining ideas and designing solutions. These practices need to be linked with the study of interdisciplinary core ideas and applied throughout students' education. Most of the standards projects that deal with mathematics suggest a greater focus on improving math achievement and identifying key ideas that determine how knowledge is organized.

Guiding Principles for Mathematics:

1. Learning: Math ideas should be explored in ways that stimulate curiosity, create enjoyment of math, and develop a depth of understanding. Students should be actively engaged in doing meaningful mathematics, discussing ideas, and applying math in interesting, thought-provoking situations.
2. Math tasks should be designed to challenge students. For example, some short- and long-term investigations can be designed to connect procedures and skills with conceptual understandings. Tasks should generate

active classroom talk, promote conjectures, and lead to an understanding of the necessity of math reasoning.

Standards for Math Practice

1. Make sense of problems and be persistent in trying to solve them.
2. Reason and draw conclusions.
3. Create reasonable arguments.
4. Model with math and science situations.
5. Use appropriate tools.
6. Be precise.
7. Clearly express your reasoning.
8. Interpret results.
9. Report on the conclusions and the reasoning behind them.

Mathematics: A Tool of Science

Although it is usually best for children to construct knowledge for themselves, we should recognize that students frequently have false understandings about math-related concepts. Some of the misconceptions that students have are natural, whereas some are picked up from the media and the home environment. Just have your students draw a picture of a mathematician or a scientist and you will have some graphic evidence of stereotypes. You might also have them keep track of representations in film and on television.

Children have a natural curiosity when it comes to using science and math to examine the natural world. They learn by experiencing things for themselves, building on what they have already learned, and talking with other students about what they are doing. Observing, classifying, measuring, and collecting are just a few examples of the processes that children of all ages can learn and apply (Leinwand, 2012).

UNDERSTANDING SCIENCE AND MATHEMATICS:

1. Science and math are methods of thinking and asking questions.

How students make plans, organize their thoughts, analyze data, and solve problems is *doing* science and math. People comfortable with science and math are often comfortable with thinking. *The research question* is the cornerstone of all investigation. It guides the learner to a variety of sources revealing previously undetected patterns. These undiscovered openings can become sources of new questions that can deepen and enhance learning and inquiry.

Questions such as "How can birds fly?" "Why is the sky blue?" or "How many?" have been asked by children throughout history. Obviously, some of their answers were wrong. But the important thing is that the children never stopped asking: they saw and wondered and sought an answer.

2. Science and math are a knowledge of patterns and relationships.

Children need to recognize the repetition of science and math concepts and make connections with ideas they know. These relationships help unify the science and math curriculum as each new concept is interwoven with former ideas. Students quickly see how a new concept is similar to or different from others already learned. For example, young students soon learn how the basic facts of addition and subtraction are interrelated ($4 + 2 = 6$ and $6 - 2 = 4$). They use their observation skills to describe, classify, compare, measure, and solve problems.

3. Science and math are tools.

Mathematics is the tool scientists and mathematicians use in their work. All of us use it daily as well. Students come to understand why they are learning the basic science and math principles and ideas that the school curriculum involves. Like mathematicians and scientists, they also will use science and mathematics tools to solve problems. They will learn that many careers and occupations require the use of science and math tools.

4. Science and math are fun (a puzzle).

Anyone that has ever worked on a puzzle or stimulating problem knows what we're talking about when we say science and math are fun. The stimulating quest for an answer prods one on toward finding a solution.

5. Science and math are art forms.

Defined by harmony and internal order, science and math need to be appreciated as art forms where everything is related and interconnected. Art is often considered subjective, and by contrast, objective science and math are often associated with memorized facts and skills. Yet the two are closely related to each other. Students need to be taught how to appreciate the scientific and mathematical beauty all around them—for example, exploring fractal instances of science and math in nature: (A *fractal* is a wispy tangled curve that seems complicated no matter how closely one examines it. The object contains more, but similar, complexity the closer one looks.). A head of broccoli is one example. If you tear off a tiny piece of the broccoli and look at how it is similar to the larger head you will soon notice that they are the same. Each piece of broccoli could be considered an individual fractal or a whole. The piece of broccoli fits the definition of *fractal* appearing complicated; one can see consistent repetitive artistic patterns.

6. Science and math are languages—means of communicating.

Science and math require the use of special terms and symbols to represent information. This unique language enhances our ability to communicate across the disciplines of technology, statistics, and other subjects. For example, a young child encountering (3 + 2 = 5) needs to have the language translated into terms he or she can understand. Language is a window into students' thinking and understanding.

Our job as teachers is to make sure students have carefully defined terms and meaningful symbols. Statisticians may use mathematical symbols that seem foreign to some of us, but after taking a statistics class, we too can decipher the mathematical language. It's no different for children. Symbolism, along with visual aids such as charts and graphs, are an effective way of expressing science and math ideas to others. Students learn not only to interpret the language of math and science, but also to *use* that knowledge.

7. Science and math are interdisciplinary.

Students work with the big ideas that connect subjects. Science and math relate to many subjects. Science and technology are the obvious choices. Literature, music, art, social studies, physical education, and just about everything else make use of science and math in some way. If you want to understand what you are reading in the newspaper, for example, you need to be able to read the charts and the graphs taught in science and math classes.

ACTIVITIES THAT HELP STUDENTS UNDERSTAND SCIENCE AND MATHEMATICS

1. *"Science and math are methods of thinking" activity:*
   List all the situations outside of school in which you used science and math during the past week.

2. *"Science and math are knowledge of patterns and relationships" activity:*
   Show all the ways fifteen objects can be sorted into four piles so that each pile has a different number of objects in it.
3. *"Science and math are tools" activity:*
   Solve these problems using the tools of science and math:
   - Will an orange sink or float in water?
   - What happens when the orange is peeled? Have groups do the experiment and explain the reasoning behind their answers.
4. *"Having fun, solving a puzzle with science and math" activity:*
   With a partner, play a game of cribbage (a card game whose object is to form combinations for points). Dominoes is another challenging game to play in groups.
5. *"Using science and math as an art" activity:*
   Have a small group of students design a fractal art picture.
6. *"Applying the language of science and math" activity:*
   Divide the class into small groups of four or five. Have the group brainstorm about what they would like to find out from other class members (favorite hobbies, TV programs, kinds of pets, and so forth). Once a topic is agreed on, have them organize and take a survey of all class members. When the data are gathered and compiled, have groups make a clear, descriptive graph showing topics of interest that can be posted in the classroom.
7. *Designing interdisciplinary activities with science and math:*
   With a group, design a song using a rhythmic format that can be sung, chanted, or rapped. The lyrics can be written and musical notation added.

LOOKING AT SCIENCE AND MATH DURING THE TWENTY-FIRST CENTURY

Science and math in the twenty-first century emphasize the processes of good practice, problem-solving, reasoning, proof, communications, making connections, and forming scientific and mathematical relationships. Digital technologies have a major role to play in these processes.

In the last century, teachers emphasized the memorization of facts and answering questions correctly. Now, more attention is paid to helping students learn how science and math relate to social problems, technology, creative innovation, and their personal lives (Peters and Stout, 2011).

A new pattern for teaching science and math is emerging that attends to the content and the characteristics of effective instruction. Engaging students in active and interactive learning deepens their involvement in their academic work and their understanding of the subjects we teach.

Good teachers know that students should have many opportunities to interpret science and math ideas and construct understandings for themselves. In the latest approaches, science and math teaching makes an effort to be purposeful and provide meaningful activities with real applications that touch people's lives. By emphasizing collaborative inquiry, promoting curiosity, and valuing students' ideas, these subjects become more accessible and interesting.

A COLLABORATIVE MODEL FOR TEACHING SCIENCE AND MATH

Views of learning emphasize thinking processes within the learner and point toward changes that need to be made in the way that educators have traditionally thought about teaching, learning, and organizing the school classroom. Central to creating such a learning environment is the desire to help individuals acquire or construct knowledge.

That knowledge is to be shared or developed—rather than held by the authority. It holds teachers to a high standard, for they must have both subject-matter knowledge and pedagogical knowledge based on an understanding of learning and child development.

The collaborative learning model for inquiry in science and math emphasizes the intrinsic benefits of learning rather than external rewards for academic performance. Lessons are introduced with statements concerning reasons for engaging in the learning task. Students are encouraged to assume responsibility for learning and evaluating their own work and the work of others. Interaction may include a discussion of the validity of explanations, search for more information, testing of various explanations, and considering the pros and cons of specific decisions.

The characteristics that distinguish new collaborative science and math learning revolve around group goals and the accompanying benefits of active group work. Instead of being told they need information, students learn to recognize when additional data is needed. They jointly seek it out, apply it, and see its power and value.

In this way, the actual use of science and math information becomes the starting point rather than being viewed as an add-on. The teacher facilitates the process, instead of acting as a knowledge dispenser. Student success is measured by performance, work samples, projects, and applications. Scientific reasoning involves testing ideas through experimentation and a creative search for new ideas and applications.

Learning science and math in today's schools has a lot to do with exploring a problem, thinking, and posing a solution. This involves peers helping each other, self-evaluation, and group support for risk-taking. This also means accepting individual differences and having positive expectations for everyone in the group.

It helps if students understand that the purpose of their tasks is to contribute to their own learning and self-development as well as the group's. If the teacher helps children push these elements together, the result will be greater persistence and more self-directed learning.

Ensuring plenty of time for active collaborative learning allows students to jointly address common topics at many levels of sophistication. This instructional method commonly involves having all students work on the same topic during a given unit. The work is divided into a number of investigatory or practical activities in which the students move from working alone to working in small groups.

Differentiated activities are organized to include basic, required work, and optional enrichment work. This way, the groups that move more slowly can accomplish the basic requirements and still be able to choose some of the options. The more able groups should move on to more challenging assignments after completing the basic tasks. Because topics are not sequenced linearly, each new topic may be addressed and differentiated instruction provided for.

OVERVIEW OF THE INTEGRATED SCIENCE AND MATH STANDARDS

All students should

- understand numbers and operations and estimate and use computational tools effectively.
- understand science and math subject matter including physical, life, and earth/space sciences, algebra, and geometry.
- understand and use various patterns and relationships.
- use observation and special reasoning to solve problems.
- understand the themes and processes of science and math.
- understand and use systems of measurement.
- become familiar with inquiry skills (pose questions, organize, and represent data).
- focus on problem-solving.
- recognize reasoning and proof as essential and powerful parts of science and math.

- communicate ideas clearly to others by organizing and using thinking skills.
- understand the relationships among science, math, and technology and make connections among them.
- identify with the history, culture, and nature of science and math.
- understand and practice science and math from a personal and social perspective.

These selected integrated standards are derived from the *National Science Education Standards* (National Academies Press, 2011).

SAMPLE ACTIVITIES

To link the integrated standards to classroom practice, a few sample activities are presented. The intent is not to prescribe an activity for a unique grade level, but to present activities that could be modified and used in many grades.

Activity 1: Solving Problems

Problem-solving should be the starting place for developing understanding. Teachers should present word problems for children to discuss and find solutions working together, without the distraction of symbols. The following activities attempt to link word problems to meaningful situations.

Objectives

Students will

- solve problems.
- work in a group.
- discuss and present their solutions.

Directions

1. Divide students into small groups (of two or three students).
2. Find a creative way to share 50¢ among four children. Explain your solution. Is it fair? How could you do it differently?
3. Whenever students collaborate to deal with problems and issues, it's best to encourage them to explain their reasoning to each other and to the rest of the class.

4. After discussing each problem, show the children the standard notation for representing division.

Activity 2: Using Statistics: Supermarket Shopping

Statistics is the science or study of data. Statistical problems require collecting, sorting, representing, analyzing, and interpreting information.

Objectives

Students will

- collect, organize, and describe data.
- construct, read, and interpret displays of data.
- formulate and solve problems that involve collecting and analyzing data.

Problem

1. Your group has $2.00 to spend at the market. What will you purchase?
2. Have groups explain and write down their choices.
3. Next, have groups collect data from all the groups in the class.
4. Graph the class results.

MAKING INTERDISCIPLINARY CONNECTIONS

Mathematics and technology have always served as important tools for work in the sciences. All three have major roles to play across the curriculum (Schwartz, 2008). Science and math inform everything from history to the evening news. They enrich the visual and performing arts, as well as sports and physical education. As an extension of our natural language, they provide a context for language learning.

As science and math continue to become more integrated into society, their interconnectedness with other school subjects becomes an important goal (Burns, 2007).

INTERDISCIPLINARY SCIENCE AND MATH ACTIVITIES

It is best to get away from the idea of splitting up the curriculum; instead, the new focus looks at fusing various disciplines together. The next few activities try to accomplish this goal.

Activity 3: Using the Sun to Teach Geometry

Objectives

Students will

- describe, model, draw, and classify shapes.
- investigate and predict the results of combining, subdividing, and changing shapes.
- develop spatial sense.
- relate geometric ideas to number and measurement ideas.
- recognize and appreciate geometry in their world.

Directions

1. Have students investigate figures and their properties through shadow geometry, exploring what happens to shapes held in front of a point of light. They can also explore what happens to shapes held in the sunlight when the sun's rays are nearly parallel.
2. Have children discover which characteristics of the shapes are maintained under varying conditions. For this activity, provide pairs of children with square objects such as wooden or plastic squares.
3. Take students to an area of the playground that has a flat surface. Have the children hold the square regions so that shadows are cast on the ground. Encourage the children to move the square regions so that the shadow changes.
4. Have students talk about the shadows they found. Discuss how they were able to make the shapes larger and smaller. See what other observations they have made.
5. To make a permanent record of shapes, have a student draw a shape on a piece of paper and put the piece of paper on the ground. Let the shadow fall on the paper. Have that student draw around the outline of the shadow.
6. When each student has had a chance to draw a favorite shape, there will be a collection of interesting drawings that can serve as a source for discussion, sorting, and display.
7. For a challenge activity, students may also wish to discover if they can make a triangle or a pentagon shadow using the square region. Using outlines children drew of the shadows cast by square shapes, see if students can find things that are alike and different in the drawings.
8. Have students count the number of corners and sides of each shape and compare those numbers. Encourage them to discuss and record their conclusions.

Precaution

Be sure you direct students not to look at the sun directly.

Activity 4: Creating Maps

Objectives

Students will

- recognize, describe, extend, and create a wide variety of patterns.
- represent and describe science and math relationships.
- explore the use of variables and open sentences to express relationships.

As students measure many geometric figures, they uncover patterns and see relationships. After students have explored different shapes, the next part of the activity will look at patterns. The topic is called *mapmaking*. A *map* is a pattern that we follow. It tells us where to go and how we go about getting there. The following activity requires students to follow a map, recording as they go.

Directions

1. Have children use graph paper and a pencil to draw a path. Each unit on the grid will represent one city block.
2. Write the labels for north, east, south, and west. Begin your map in the middle of the graph paper. Then, follow this route:
   A. Walk two blocks south.
   B. Turn east and walk three blocks.
   C. Turn south and walk one block.
   D. Turn east and walk three blocks.
   E. Turn south and walk four blocks.
   F. Turn east and walk one block.
   G. Turn south and walk three blocks.
   H. Turn west and walk half a block.
3. Compare your map with those of other students. How are they the same? How are they different? How would the map change if step B were a 90-degree turn west?
4. Challenge problem: With your group, make a scale model of the local area. Include "key" identifying symbols and directions.

Activity 5: Estimate and Weigh Different Materials

Objectives

Students will

- estimate the weight of different objects.
- relate everyday experiences to the science/math measuring activity.
- check their estimates.

Through many different experiences in representing quantities and shapes, children establish a link between their concrete everyday experiences and their understanding of science and math abstraction. Representation helps children remember an experience and make sense by communicating it to others.

Directions

1. Fill several milk cartons with different materials such as rice, beans, clay, plaster of Paris, and wooden objects.
2. Seal the cartons and label them by color or letter. Tell the children what the materials are, but do not identify the contents of a particular carton.
3. Have the children guess how to order the cartons by weight according to what they contain. Then, let the children order the cartons by weight, holding them in their hands, and using the balance scale to check their estimates.

Activity 6: Which Will Melt Quicker?

Objectives

Students will

- explore reasoning in science and math.
- measure and compare quantities.
- write about their findings in their science and math journals.

Problem

Suppose you have a glass of water. It has the same temperature as the air. Would an ice cube melt faster in the water or air? Invite students to find out.

Materials

thermometer, water, ice cubes, two identical drinking glasses, small plastic bag, salt, spoon

Directions

1. Fill one container with warm water and leave the other container empty.
2. Let the children see and feel the cups and the water. Explain, "We are going to put an ice cube in each container. Our problem is this: Which ice cube do you think will melt first?" Write the guesses on the board.
3. Have students measure the temperature inside the empty glass. Also, measure the temperature inside a glass of water. It should be about the same as the air temperature. If not, let the water stand a while.
4. Find two ice cubes of about the same size.
5. Put one cube into the empty glass. Put the other in the glass of water.
6. Compare how fast the ice cubes melt.

Questions for Further Investigation

1. How can you make an ice cube melt faster in water? Will stirring the water make a difference? Will it melt faster in warmer water? Does crushing the ice make a difference? Does changing the volume of the water matter?
2. How will ice cubes melt when other things are added to the water?
3. Will an ice cube melt faster in salt water? Does the amount of salt make a difference?
4. Students will measure and compare temperatures, hypothesize, experiment, and arrive at conclusions.

Evaluation

Have students explain their reasoning through writing about their experiment in their journals. Direct their discussion by asking them to explain what mathematics they used. What science skills were involved? What is the best way to show their data?

Activity 7: Design an Escher-Type Art Drawing

Students become aware of the properties of shapes through many experiences. They manipulate, visualize, draw, construct, and represent shapes in a variety of interesting, creative ways.

Objectives

Students will

- create patterns.
- explain their designs.

Directions

1. Introduce the idea of making a repeated pattern—*tessellation*. Teachers may wish to talk about artists who use the concept of tessellation in their artwork.
2. Have students represent a three-dimensional object on paper.
3. Ask students what shapes can be seen in different objects.
4. Have them try to make a symmetrical design. What shapes will tile a floor (or tessellation)?
5. Begin by having students make a tessellated design. Encourage students to explain their pattern and the relationships between the figures they've chosen. Students who are adept at tessellating (drawing repeated patterns) may wish to extend that skill in artwork designs.

Through direct experiences with three-dimensional objects and then transferring those objects into a two-dimensional world, students become aware of the relationships and properties of geometric shapes. They are able to notice symmetry in the designs they create. They investigate how patterns look if they're moved or rotated. They draw, build, and describe many shapes from a variety of perspectives.

Activity 8: Finding Your Heart Rate

Background Information

The heart is a pump. The heart pumps blood to parts of our body. The number of times the heart pumps per minute is called *heart rate*. Heart rate changes when we do different activities.

Objectives

Students will

- understand the heart and its system.
- learn how to calculate their heart rate.

- understand that heart-rate changes depend on physical activity.
- chart heart-rate data.

Materials

Classroom with room to move around, calculate-your-heart-rate worksheet, pencils, timer or clock

Procedure

Introduce *heart rate* and have students guess how many beats per minute their heart is beating. Ask students to name activities they believe would change heart rate. Then, ask if they think these activities would make the heart beat faster or slower than normal. Explain to students that there are two main areas on the body where it is easiest to find your heart rate (pulse at the neck and wrist). Have students find their pulse. Make sure each student has found it. Practice counting while being timed for ten seconds.

Directions

1. Pass out the heart-rate worksheet. Have students count their heart rate while sitting.
2. Have students enter the number on a chart. Explain to students that this is how many times the heart beats in 10 seconds.
3. Explain that the heart rate is taken in one-minute intervals. Help students multiply their number by 6 in order to have the number of times their heart beats in one minute. Enter the number on the worksheet.
4. Do the same for activities of standing and running in place for 30 seconds.
5. After the worksheet chart is filled in, ask students to write a sentence about a time when they felt their heart rate change.

Evaluation

Students' performance will be evaluated according to how much of the chart they are able to fill in and how well they participated in the activity.

Heart-Rate Worksheet

Your heart rate is how many times your heart beats per minute.
Your heart rate changes.
Your heart rate is your pulse.
Your pulse is found in your neck and wrist.

Record your heart rate below:

	Sitting	Standing	Running in Place
10 seconds	_____	_____	_____
30 seconds	_____	_____	_____
60 seconds	_____	_____	_____

A useful instructional procedure is to determine the purpose and scope of the lesson. Next, build on student interests as you provide space for thinking, reflection, and discussion. Be prepared to push student thinking forward with purposeful activities. Finally, share some of your lessons with other teachers and get suggestions for improvement.

Whether it is individual or group work, it also makes sense for inquiry, scientific reasoning, and mathematical problem-solving skills to be integrated and utilized across the curriculum.

A SAMPLE OF ONLINE RESOURCES FOR STEM

Connected science, technology, engineering, and math activities and projects are part of today's science and math curriculum. The learning process associated with STEM views science and math as underlying all engineering problems. Technology is viewed as an essential tool in the search for answers.

Organizations like the *National Science Teachers Association* (www.nsta.org) and the *National Council of Teachers of Mathematics* (www.nctm.org) can help. So can the following websites:

- *AAAS Science NetLinks* (http://sciencenetlinks.com)
  The American Association for the Advancement of Science has made available a wide range of lesson plans for just about every grade level.
- *Discovery Education* (www.discoveryeducation.com/teachers)
  Here the content fits in nicely with state standards and there are links to a large number of teacher friendly math websites.
- *The National Science Digital Library (NSDL) List of Resources of Physical Sciences for the Lower Primary Grades* (https://nsdl.oercommons.org/browse?batch_size=20&sort_by=title&view_mode=summary&f.sublevel=lower-primary&f.general_subject=physical-science)
  This site gives teachers a wide variety of science teaching resources for the lower primary level before they have to teach them. Also, there are a wide range of short lessons—covering everything from 2013 phases of the moon to a three-dimensional model of the Big Dipper.
- *Bring the World to Your Classroom* (https://whyy.pbslearningmedia.org)

Here, teachers and students can choose from thousands of resources—including lesson plans, videos, and interactive activities.
- *NASA's Search for Another Earth* (https://exoplanets.nasa.gov/)

If students or teachers want to connect to experts, this is a good place to do it. Possibilities range from NASA's Jet Propulsion Laboratory (JPL) and images from space to the "Ask an Astronomer" podcast. Other possibilities include online games, activities, and submitting questions to experts.

All of the sites referred to here invite curiosity in a way that stretches students' minds. Some of the visuals instill a sense of wonder while many of the activities engage learners in a way that helps them ask insightful questions.

CHALLENGING SCIENCE AND MATH ACTIVITIES

This section brings together the science and math standards to elementary and middle school classrooms. Meaningful activities that employ the core ideas and practices of observing, comparing, measuring, recording data, and making good conclusions are emphasized. These activities are based on the 2010 National Framework for K–12 Science Education Standards (National Academies of Sciences, Engineering, and Medicine, 2012). Whenever possible, the Mathematics Standards (National Council of Teachers of Mathematics, 2018) are included.

Activity 1: Mysterious Stories

Stories are a means of communication. Even before written language, people drew pictures on the walls of their caves to show a successful hunt or the animals they met. Stories told of rules and accomplishments. Whether stories are found in dance, pictures, art, or words, it is important to view them as a desire of people to communicate their thoughts, dreams, and mysteries across generations. Looking at stories as mysteries is exciting, providing students with characters they can identify with, which allows them to be included and be part of the adventure.

Activity 2: Invent a Mystery Story

Create your mystery story.

Activity 3: The Mystery of Gravity and Yo-Yos

Grade level: elementary and junior high

Content Standards

- Applying the core ideas of physical science
- Comparing movement and gravity

Background Information

Students should review the basic physics principles of gravity and inertia.

Objectives

Students observe objects concerning force, acceleration, friction, and gravity. Students apply physics to real-world situations.

- gravity: a natural force of attraction that tends to draw objects together
- inertia: a property of matter whereby it remains at rest or continues in uniform motion unless attracted by some outside force
- velocity: rate of change of an object's position

Questions

How do yo-yos reflect physics?
What is an example of inertia?
How does your yo-yo show force?
Can you make your yo-yo accelerate?
What causes friction when using a yo-yo?
Can you explain why a yo-yo shows gravity?

Other Student Activities

Students work in pairs with the yo-yos.

1. Estimate the direction your yo-yo moved in one minute.
2. Record the velocity of your yo-yos. At what speed do you think they traveled?
3. Write a mystery story about your yo-yos and how they demonstrate gravity.
4. Why does a yo-yo act as it does?
5. There are some professional yo-yo groups. Find out more about them. Record your findings.

Accelerated Student Activities

1. Slow down your yo-yo, then, try to speed it up.
2. What happens when the yo-yo is thrown in a different direction?
3. What makes the yo-yo slow down?
4. When does the yo-yo move fast?
5. Describe the physics involved in these activities.
6. Record your yo-yo movements. Compare your findings with those of other students.
7. With other students, describe the force, speed, gravity, and friction of your yo-yos. Record your findings.
8. Make a class chart so that other classes can see your physical science achievements,

Activity 4: Flowing Mysteries

Grade level: elementary and junior high

Content Standards

- Applying the physical science standards
- Using basic ideas and procedures in science
- Putting personal views to practical use
- Practicing oral and written communication skills
- Employing science and technology skills

Objectives

Students experiment with household substances.

Materials: Student Tools for Each Work Station

6 small plastic bottles
6 flowing substances (vinegar, soap, alcohol, cooking oil, vanilla, water)
6 short glass tubes with rubber bulb
1 wide-mouthed drinking cup
1 small drinking cup
1 flat-bottomed container for holding articles
1 piece of Saran Wrap
1 sheet of aluminum foil
1 sheet of waxed paper
1 sheet of white paper

Directions

1. Prepare the containers with food coloring added.
2. Have students discuss how scientists perform experiments.
3. Prepare the trays and tools for each group.

On the chalkboard, list the experiments that students might try:

- substance races
- floating ability
- density
- combining liquids
- other suggestions

Student Task

Discover what six substances are using the following rules:

- Use your sense of sight to find out what the flowing mysteries are.
- You are not to smell, touch, or taste the substances.
- Each dropper can be used to pick up only one substance,

Students rotate among the work stations, experimenting as they try to discover what the flowing mysteries are. They conduct several tests during the process.

Work Station 1: Substance Race

Lesson steps:

1. Choose a substance and a sheet of paper to cover your tray (waxed, white, aluminum foil, Saran Wrap).
2. Place a drop of each substance on the paper.
3. Tip the tray so the substances move.
4. Record the movement of the substances.
5. Try the experiment with all of the flowing substances.

Work Station 2: Floating Ability

Lesson steps:

1. Select a small plastic container.
2. Add drops of each colored substance.
3. We want to see which substances will float.

4. Encourage students to experiment.
5. Jerk, shake, and maneuver the container to detect which substances move to the top.
6. Record the movement of the substances.

Work Station 3: Density

1. Select a small plastic container.
2. Add drops of each colored substance.
3. We want to see which substances will sink.
4. Encourage students to experiment.
5. Jerk, shake, and maneuver the container to detect which substances sink.
6. Record the movement of the substances.

Work Station 4: Mixing Substances

1. Guess what each substance is and test your guesses by observing which substances blend together.

Record your guesses.

blue substance
green substance
red substance
yellow substance
purple substance
clear substance

Which will mix? Write your reasoning:

vinegar
soap
alcohol
cooking oil
vanilla
water

CONNECTING THE MATH STANDARDS TO THE CORE CONTENT STANDARDS

Standard 1: Understand Number and Operations

Students need to understand counting (represent one-to-one correspondence with concrete materials, match a set to a numeral).

Standard 2: Ability to Use Patterns, Algebra, Functions, and Variables

Students will understand and use functions (plus +, minus –, times x, divide ÷).

Standard 3: Geometry

Apply geometry; understand shapes; use size, symmetry, congruence, and similarity.

Standard 4: Measurement

Use measurement to measure and compare lengths and widths; tell and write time in hours and half hours; use analog and digital clocks.

Standard 5: Data Analysis, Probability

Organize data; use charts, tables, graphs, and statistics to make sense out of math.

Standard 6: Problem-Solving

Find solutions, use strategies, take risks, make decisions, and get results.

Standard 7: Reasoning

Reasoning is connected to students' language development. Thinking and reasoning are important to math learning. Students should have experiences with deductive reasoning (moving from guesses to conclusions).

Learners should also be aware of inductive reasoning (informal reasoning using specific examples) and use evidence to make assumptions and form conclusions.

Standard 8: Communicating

This includes working in groups talking, listening, and expressing ideas. Students share information, explain ideas, and help each other.

Standard 9: Forming Conclusions

Many relationships are learned every day by students making connections through their own experiences and applying math content to real-life situations.

Standard 10: Representing Math Relationships

Representing zero (0) is showing connections among math concepts to improve understanding. When students are able to use different blocks, colored math squares, and fraction pieces while performing math skills, they are more likely to enjoy their lesson.

INTEGRATING SCIENCE, MATH, AND ARTS ACTIVITIES

We need all these tools—science, math, the arts, literature, and more—not only to fully express ourselves, but also to explore serendipitous connections that lead to creativity and innovation. Imaginative new ideas and products are rarely completely original. In fact, innovation is often the result of a step-by-step process that builds on ideas that are at least partly developed by others. The team that puts the puzzle together in a unique way—*and acts on the results*—usually gets the credit.

This STEM lesson contains elements from many sources, including ideas from Atlantic Canada's science curriculum that deals with sound, engineering, and music.

Activity 1: Song Lyrics

Materials

Each group of two or three gets the following:

- scissors and balloons
- duck and masking tape of different colors and designs
- different kinds of string, sewing thread, rubber bands, and plastic cups
- a box about twelve-by-nine inches in size and about three inches deep. (We got ours from Staples where they use a good-sized box whenever you get something copied.)

Procedure

Divide the class into groups of two or three. Have one small group practice a song using lyrics that are printed on a piece of paper. Practice when no one but the singers are in the room; no memorization. If you can play the guitar (or if there is a piano in the room), you can do it. If not, see if anyone in the room can play the guitar or the piano; be sure that they are in the practice group. If no one can play any instrument, just have the song playing in the background as three or four students sing.

Instrument-Building Activity (after the Song)

The group starts by putting their design for an instrument on paper.

Give students time to build and try out their instrument.
Use the same song you used to start the lesson; each group should use their instrument as the song is sung by the entire class. Do the same thing a second time if you think that is appropriate.
How many different sounds can your instrument make?
Explore the science of sound.

Activity 2: The Super Creature: Structural Features of an Animal that Enable It to Survive in Its Habitat

NOTE: This STEM lesson was designed by Peter Cudmore.

This lesson is intended for grade three and up. It is designed to encourage students to use their knowledge of science and math to engineer features needed for an animal to survive in multiple habitats.

Objectives

1. Students will compare the external features of several organisms and relate these features to the basic survival of the animal in its natural habitat.
2. Students will predict the structural adaptations needed for an animal to live in a particular habitat—real or imagined.
3. Students will work in groups of three or four to choose an animal they wish to adopt.
4. Students will draw and present their evolved organism's features and explain how it can live in three of the six given habitats.

Materials

Markers of different colors, pencils, butcher paper (2.5 feet by 4 feet), and any item students need to manipulate their work.

Procedure

1. The teacher will ask for examples of special features that help animals survive and attain their basic needs (food, water, shelter, and air).
2. More examples of special features will be presented using pictures and characteristics explained by the teacher.

3. The teacher will describe the task of building an animal with special features that allow it to adapt to at least three different habitats. Six different habitats will be given by the teacher with pictures and a written description of each.
4. Students will be divided into groups of three or four to collect materials to complete their project.
5. Students work in their group to create adaptations to their chosen animal. Adaptations must be noted and labeled. These evolved animals will be presented to the class.

Evaluation

Together, groups will present their animal and how it can survive in three of the six given habitats. Students must explain why each new adaptation will benefit their organism.

Activity 3: Fantastic Friction

Curriculum Standards and Outcomes

- Predict which factors will affect the motion of an object.
- Using terms like "faster" and "slower" and tools such as rulers, string, and stopwatches to test predictions.
- Drawing simple conclusions about the factors that affect movement based on their investigations.

Background Information

This experience is designed to help children collaboratively explore the force of *friction*—which is the force that is created whenever two surfaces move or try to move across each other. The amount of friction depends on the texture of both surfaces and the amount of contact force that is pushing the two surfaces together. Students will have the opportunity to develop a sense of space, orientation, perspective, and the relationship between different objects and materials.

Objectives

To teach students that motion relies on the "rolling resistance" of a specific surface.

To help students understand that:

- Friction always opposes the motion or attempted motion of one surface across another surface.

- Friction is dependent on the texture of both surfaces.
- Friction is also dependent on the amount of contact force pushing the two surfaces together.

Materials

- Wooden ramp
- Toy cars
- Piece of carpet
- Piece of plastic
- Piece of Velcro
- Piece of hardwood
- Metric and yard stick
- Stickers

Procedure

Introduce the lesson with the "Bill Nye the Science Guy" video *Friction*. You may want to view most of the video in class—or view a few clips in the entire class and let the students know where they can get to see the whole segment as homework. Either way, this approach works for visual and oral learners as children listen, watch, and discuss. After the video, ask the children for some important words to define and discuss. Examples include rub, float, stick, faster, slower, and stop.

If you can't get the video for the classroom, just cover the words and ideas yourself.

Activity

(NOTE: This STEM lesson was designed by Sarah MacKay.)

In groups—or in the whole class—it is possible to do an experiment involving a ramp, toy cars, and four different surface materials. The basic idea is to test different surface materials to predict which one will produce the most and the least amount of friction. Try to draw some scientific conclusions from the observations.

You need a wooden ramp about eight inches wide and about forty inches long. Try to get it at about a 45-degree angle. Place four different types of surface materials at the bottom of the ramp, each with a different level of friction. Predict how far a toy car might go. As a car exits down the ramp, measure the distance the car goes on different surface materials. Mark where the car stops on each material and decide as a group or as a class which material produces the most and the least amount of friction. Students compare predictions with what actually happens.

Assessment:

Students will each be given a worksheet to record their estimations and eventual findings after the experiment. Next, they will compare their estimates and their findings and arrive at a conclusion. Look for key words such as "faster," "slower," "fastest," "slowest," and so forth.

Finish with an overview of the activity as a class and reflect on the answers together.

Challenges are more than puzzles to be solved. When you start, some of the pieces may not even be on the table. In the process of working toward solving a problem, new things come up and change things. So it's best to leave room for the unexpected.

SUMMARY, CONCLUSION, AND LOOKING FORWARD

The insights of modern science have come about with the help of mathematics and technology. Combining these tools with the processes of scientific reasoning opens up possibilities for discoveries that lead to a better understanding of the world.

The imaginative implications of science, math, and their technological associates reach well beyond the schools. Mastering these subjects may not result in innovation, but they certainly increase the odds of it happening.

As far as mathematics instruction is concerned, teachers have to go beyond teaching students basic arithmetic, how to balance a checkbook, or estimate how long it will take to get from one town to another. Students should learn to ask the big questions—as they consider how math might be applied to broader human problems.

As far as specific classroom content is concerned, it is not enough to teach computation and procedures in isolation from the situations that require those skills. Those who teach science and math also have a responsibility to meet the challenges of technology and society.

To thrive creatively in the classroom, it helps if students can see the unique work of others. And they need to be in an environment where imaginative work is valued. Cognitive capacity and subject-matter knowledge matter, but creative individuals tend to have risk-taking personalities and temperaments. Still, like the idea of multiple intelligences, there are many different kinds of creativity—and there are multiple paths to imaginative ideas (Sinclair, 2006).

Whatever approach you take incorporating creativity into your lessons, it is important to remember that failure is widely recognized as part of the creative process (Weld, 2017). Still, it certainly helps if you fail quickly and know how to learn from false starts and mistakes.

We all know that every idea generated is not going to be a good idea. However, the more new ideas that are produced, the more likely it is that something unique and useful will turn up.

Being naive or afraid of science or math can be a real problem in school, in the workplace, and in a democratic society as a whole. The key to academic success in these subjects is fostering habits of the mind such as critical thinking, problem-solving, agility, adaptability, curiosity, and imagination. Also, in a world filled with the technological products of science and math, it is more important than ever to understand these subjects and how they connect to our daily existence.

The need for analytical skills and social intelligence goes with any version of the future. But, in spite of certain agreed-upon principles, uncertainty and change go with the territory. While it is true that an adventure into the unknowable future is often feared, one of the certain lessons of history is that everything changes.

Twenty-first-century skills include having some idea about how to use the intellectual tools of science and math in purposeful and meaningful ways. A good way to get ready for the situations and big ideas that are just beyond the horizon is to look around today. Notice how the future often gets mashed up with the past.

In any discussion of the future, it is important to keep in mind the caution that history doesn't repeat—it rhymes.

> The future is already here. It's just not evenly distributed yet.
>
> —William Gibson

REFERENCES

Andrews, L. (2012). *I Know Who You Are and I Saw What You Did*. New York: Free Press.

Burns, M. (2007). *About Teaching Mathematics: A K–8 Resource*. Sausalito, CA: Math Solutions.

Cain, S. (2012). *Quiet: The Power of Introverts in a World That Can't Stop Talking*. New York: Crown Publishers.

Dweck, C. (2006). *Mindset: The New Psychology of Success*. New York: Random House.

Eichinger, J. (2004). *40 Strategies for Integrating Science and Mathematics Instruction: K–8*. Upper Saddle River, NJ: Prentice Hall.

Gibson, W. (2012). *Distrust That Particular Flavor*. New York: G. P. Putnam's Sons.

Jadrich, J., and Bruxuoort, C. (2011). *Learning and Teaching Scientific Inquiry: Research and Applications*. Arlington, VA: NSTA Press.

Lannin, J., Ellis, A., Elliot, R., and Zbiek, R. (2011). *Developing Essential Understanding of Mathematical Reasoning For Teaching Mathematics in Grades PreK–8*. Reston, VA: National Council of Teachers of Mathematics (NCTM).

Leinwand, S. (2012). *Sensible Mathematics: A Guide for School Leaders in the Era of Common Core Standards*. 2nd edition. Portsmouth NH: Heinemann.

Lindquist, M., and Clements, D. (March 2001). "Principles and Standards (2001)." Reston, VA: National Council of Teachers of Mathematics (NCTM).

National Academies of Sciences, Engineering, and Medicine. (2012). *A Framework for K–12 Science Education: Practices, Crosscutting Concepts, and Core Ideas*. Washington DC: National Academies Press.

National Academies Press. (2011). *National Science Education Standards*. Washington, DC: National Academies Press.

National Committee on Science Education Standards and Assessment. (1996). *National Science Education Standards*. Washington, DC: National Academies Press.

National Council of Teachers of Mathematics (NCTM). (2010). *Principles and Standards for School Mathematics*. Reston, VA: National Council of Teachers of Mathematics.

National Council of Teachers of Mathematics (no date). The Core Standards for Mathematics. Reston, VA: NCTM.

Neagoy, M. (2012). *Planting the Seeds of Algebra, Pre-K–2: Explorations for the Early Grades*. Thousand Oaks, CA: Corwin.

Peters, J., and Stout, D. (2011). *Science in Elementary Education: Methods, Concepts, and Inquiries*. 11th edition. Boston: Allyn and Bacon.

Russell, S., Schifter, D., and Bastable, V. (2012). *Connecting Arithmetic to Algebra: Strategies for Building Algebraic Thinking in the Elementary Grades*. Portsmouth, NH: Heinemann.

Schwartz, S. (2008). *Teaching Young Children Mathematics*. Lanham, MD: Rowman & Littlefield.

Shtulman, A. (2017). *Scienceblind: Why Our Intuitive Theories about the World Are So Often Wrong*. New York: Basic Books.

Sinclair, N. (2006). *Mathematics and Beauty: Aesthetic Approaches to Teaching Children*. New York: Teachers College Press.

Small, M. (2008). *Making Math Meaningful to Canadian Students, K–8*. Toronto: Nelson Education.

Smith, M., and Stein, M. (2018). *5 Practices for Orchestrating Productive Mathematical Discussion*. 2nd edition. Thousand Oaks, CA: Corwin.

Wedekind, K. (2011). *Math Exchanges: Guiding Young Mathematicians in Small Group Meetings*. Portland, ME: Stenhouse Publishers.

Weld, J. (2017). *Creating a STEM Culture for Teaching and Learning*. Arlington, VA: National Science Teachers Association (NSTA) Press.

Witzel, D. (2012). *Teaching Science and Math: Resources and Strategies for K–12 Science and Math Teachers*. Arlington, VA: National Science Teachers Association (NSTA) Press.

Chapter 5

Language and Literacy
Communications Skills in a Digital Age

> We stand on the verge of a second great explosion—the written word is poised to change once again. It is education not technology that will ensure the future of language and literature.
>
> —Martin Puchner

Language skills provide the foundation for accessing all knowledge. Instruction in the language arts provides students with many of the intellectual tools needed to understand subjects across the curriculum.

Learning various communication skills can increase student curiosity and provide them with the skills to develop their knowledge, imagination, and power. This all goes hand in hand with exploring new ideas and creating something new and different. There is general agreement that without language and literacy skills, students will have extreme difficulty exploring concepts across the curriculum.

Success in all subjects is intertwined with competency in the language arts. A broad experiential knowledge base is one of the keys to fluency. In reading, for example, comprehension is built on what students already know. When cognitive, creative, and collaborative skills are added to the mix, every student has a chance to develop their potential for lifelong learning and participating in society as a whole.

Building student ability upon communication skills opens up the possibility of going on to create and innovate (Costigan, 2018). But a note of concern: reading prose on a screen is rapidly being eclipsed by the power of audio and video. The printed word on paper is not going away. Still, having an online culture where text is being overwhelmed by sound and images can be overwhelming.

TEAMWORK AND CREATIVE COMPETENCY

Common elements in successful classes and successful schools are a socially integrating sense of purpose and a shared sense of community. By building on the social nature of learning, teachers can encourage students to believe that they have the ability to learn. Hopefully, along the way, learners will realize that by making a real effort, they can maximize their potential.

Invention and discovery are amplified in language-rich classrooms as learners are encouraged to express themselves in unique ways. Young people soon realize that reading, writing, and other language skills help them discover and create something new. New ideas and creative solutions require more than smarts; they call for imagination and curiosity. It's also more about hard work than lightning bolts full of insights coming out of nowhere.

It is important to recognize the fact that language and literacy development depend on the quality of what students read—or have read to them. Reading strongly influences writing—and both can be integrated with the other language arts to reach across the curriculum. It should also be noted that, whether it is at home or at school, the quality of the words adults use is as important as the quantity of words used.

By the time children reach the upper elementary grades, teachers often use digital technology to enhance the development of communication skills. Screens and machines can help and/or hinder the development of reading, writing, speaking, and imaginative skills. Yes, new media may be able to help with old problems, but they can also create some new ones. Texting and multitasking during class, for example, will destroy the best lesson.

Today's cloud-based information-technology structures are bound to help us refine the information and language that we use. Along the literacy glide path, our tech tools are bound to strongly influence the way we discover and create things.

Smart machines and artificial intelligence are increasingly shaping how information is collected and used. Not only that—but now, we have robo-writers that can go beyond reprocessing data to creating human-sounding printed pages. So, it's hard to avoid the fact that new media will increasingly influence how children and young adults construct meaning with language and literacy.

As far as innovation itself is concerned, the process is less about solo inventors than about collaboration. In the right environment, the ability to work in teams can make individuals even more creative and empowered (Catmull, 2014). Add the right mix of supportive social and cultural factors to the mix and you can create something very human and very special.

It is often in the spaces between individual differences that imaginative group discoveries are made and new insights created. Examples of collaborative work include parallel computation technologies, better algorithms, and big-data collection. Taken together, these developments mean that we had all better get used to a higher level of predictive strength.

A note of concern: the technology may be helpful in predicting what we want online and even foresee a few over-the-horizon events, but we better move quickly to fine-tune the technology before it fine-tunes us.

ADVANCING LITERACY IN A NEW AGE

With or without high-tech tools, programs with connected language and literature components provide a solid foundation for approaching language instruction in a natural and holistic manner. Knowing the words is one thing, but understanding them in context is quite another (Johnson, 2014).

Comprehension is enhanced by rich experiences, a broad vocabulary, and a solid base of factual knowledge. But wherever the students are on the continuum, it is up to the teacher to make sure that young people learn to apply a wide range of strategies for comprehending, critiquing, and appreciating written, spoken, and visual language.

How might the Web and its digital associates have a detrimental or positive effect on language and literacy development? Whatever your answer, becoming too dependent on digital technology isn't a good idea.

No one knows for sure what will crop up just around the corner; but when it comes to education, diminishing the face-to-face human factor can undermine both the individual and the society. Good literature and collaborative, small-group work are and will continue to be two of the keys to advancing literacy (Taylor and Duke, 2014).

As learners move through the stages of language learning, discussions about what's being read and written helps everyone move toward fluency. The thoughtful use of all available tools can help accomplish shared language, literacy, and creative-thinking goals. What about new ideas? Coming up with them has a lot to do with remixing the metaphors, concepts, and scientific understandings of our time.

Certain simple truths about learning are sometimes lost in the smoky quarrelsomeness of the debate over the best approach for teaching. One of these simple truths is that language learning is social and instruction is most effective when it is taught holistically and within a meaningful context. Another reality is that integrating the language arts is a powerful way to connect students to the full range of real communication possibilities.

As language and literacy instruction increasingly expands to include electronic media, it is important to realize that each communications medium relates directly or indirectly to every other. Language learning, at its best, involves becoming active, critical, and creative users of print, spoken language, and the visual language of electronic media (Moline, 2012).

Technological resources can do many things—including linking the classroom to the outside world in ways that extend the boundaries of learning. Still, digital technology has a mixed track record when it comes to improving reading and writing skills at the elementary school level (Johnson, 2014).

Whiz-bang technology, collaborative learning, and a standards-based curriculum all help. But the key to high-quality instruction is a cooperative language-rich classroom where the teacher knows the characteristics of effective instruction. Meaningful group activities and settings can help students learn how to use language to communicate, solve problems, and meet the diverse literacy demands that they will encounter throughout their lives.

As teachers look for new ways to make the language arts more active, dynamic, purposeful, and fun, it is important to recognize the importance of personal adaptability and creativity. Just as there is no unified educational theory that explains everything, no instructional method will meet the needs of all students all the time (Serravallo, 2018).

Rigid scripts designed by others have always been a poor substitute for well-educated teachers who can combine professional flexibility with a thorough knowledge of effective instruction. Permanent competency in any field is elusive. The best advice for both teachers and students is this: do the best you can with what you know today—and work hard to do better tomorrow.

MAKING CONNECTIONS IN GROUPS

> Language learners must invent and try out the rules of language for themselves through social interaction as they move toward control of language-for-meaning.
>
> —G. Pace

Whether its old or new media, the whole range of language skills can be helped by communicating with peers who provide immediate feedback during the reading, writing, and revision process. It should come as no surprise to learn that the most competent readers tend to be the most competent talkers, listeners, writers, viewers, and thinkers.

Let's look at how younger children can share *big books:*

Steps in Sharing a Big Book

- The teacher introduces the book and has children predict what it is about.
- The teacher reads the book to the students, holding it so children can see words.
- The teacher sometimes pauses so students can think what will happen next.
- The teacher rereads the book; students can read along (out loud) with him or her.
- Students may pair up for a rereading, taking turns reading small parts.

A beginning reader may have to be repeatedly exposed to a book by hearing the story read well several times before they develop oral fluency themselves. The old-fashioned, round-robin style doesn't work well. Just about anything else will get the job done. If you want to do a lot of oral reading, simply do paired reading in groups of two. Get close, point the chairs in opposite directions, and take turns reading.

Don't assume that a student—or even an adult—can make a book sound exciting when reading out loud. Many new teachers have to practice and get suggestions from their peers to do the job well. So you know that young readers will have to go over the text several times with a friend or two.

Just about any activity that you can think of can be employed. For example, children can work in small groups to act out a story or practice reading the lines in a reader's theater activity. They can also work in small groups to discuss the theme, plot, characterization, or difficulty to be overcome in the story. Felt pens and large paper can be used for semantic maps or webs showing their answers. These can be put up and shared with the whole class. The most important thing is to get children thinking, discussing, and interacting with literature.

> Reading is thinking with the mind of a stranger.
>
> —Jorge Luis Borges

SOCIAL PROCESSING AND THINKING ABOUT WRITTEN EXPRESSION

The development of a writing community is a very powerful way for students to collaborate in developing their writing voice (Willingham, 2017). Whether writing is self, peer, or teacher evaluated, it is important not to lose sight of the connection between what is valued and what is valuable. Jointly developed folders (portfolios) have a major role to play in student-writing assessment. By selecting samples, these folders can provide a running record of students' interests and what they can and can't do.

To work toward less control, teachers need to help students take more responsibility for their own learning. The ability to evaluate does not come easily at first, and peer writing groups will need teacher-developed strategies to help them process what they have learned. The ability to reflect on being a member of a peer writing team is a form of metacognition—learning to think about thinking.

The skills of productive group work may have to be made explicit. This requires processing in a circular or U-shaped group where all students can see each other. Questions for evaluative social processing might include the following:

- How did group leadership evolve?
- Was it easy to get started?
- How would you feel if one of your ideas were left out?
- What would you do if most members of your group thought that you should have written something differently?
- How did you rewrite?
- Did your paper say what you wanted it to?
- What kind of a setting do you like for writing?
- How can you arrange yourself in the classroom to make the writing process better?
- What writing tools did you use?
- How do you feel when you write?
- Explain the reasoning behind what you did?

Remember, it's just as important for students to write down their reasoning as it is to explain their feelings and preferences.

The recognition of developmental stages in social skills must be taken into account as teachers incorporate literature-based writing concerns into their classroom routines. For the younger students, the writing process can take the form of jointly produced language experience stories. These stories can be placed on large charts with the teacher or an upper-grade student doing the writing. As soon as they can write on their own, the children can keep a private journal where they label drawings, experiences, and writing samples.

As students learn to expand their perspectives, they can begin to carry a story from one page (or day) to the next. Time may be set aside each day for a personal journal entry. Although it's important that the language be in a student's own words, the teacher can make comments without formal grading (Dierking and Jones, 2014).

There are times when teachers have to intervene to assess students' writing or do some final editing before something is widely shared. Remember to stamp "draft" or "creative writing—work in progress" on anything that might go home before it reaches its final form. A "work in progress" stamp would save you from a little embarrassment when a misspelled word or some bad grammar reaches parents.

STRUCTURED AND UNSTRUCTURED POETRY EXPERIENCES

Some twenty-first-century Americans, for example, might disagree with Wordsworth, a nineteenth-century Englishman who explained poetry as "emotion recollected in tranquility." People often reach for poetry today because, in its own peculiar way, poetry tells truths that other communication techniques often miss.

Teachers must have some basic knowledge of the vocabulary of poetry in order to help children enjoy and mature in their understanding and appreciation of it. One thing teachers can do is to share some of their efforts at poetry with their students—as well as reading published poetry to their class. Reading or writing poetry involves awareness of certain elements that make it unique. Some poetry characteristics that students should know about include these:

1. Poetry uses condensed language so *every* word becomes important.
2. Poetry uses figurative language (e.g., metaphor, simile, personification, *irony*).
3. The language of poetry is often rhythmical (regular, irregular, metered).
4. Some words may be rhymed (internal, end of line, run-over) or non-rhyming.
5. Poetry uses the language of sounds (alliteration, assonance, repetition).
6. The units of organization are line arrangements in stanzas or idea arrangements in story, balance, contrast, build-up, or surprise.
7. Poetry uses the language of imagery (sense perceptions reproduced in the mind).

DIFFERENT KINDS OF POETRY

1. Fixed Forms

a. Narrative or storytelling
b. Literary forms with prescribed structures (e.g., limerick, ballad, sonnet, haiku, others)
c. Lyric

2. Free Verse

a. Tone: humorous, serious, nonsensical, sentimental, dramatic, didactic
b. Content: humor, nonsense, everyday things, animals, seasons, family, fantasy, people, feelings, adventure, moods
c. Time of writing: contemporary, traditional

3. An Example of Collaborative Poetry

Students work in small collaborative groups. Each team or partnership is given a short time (one or two minutes) to compose the first line of a poem. On a signal from the teacher, each team passes their paper to the next group and receives one from another. The group reads the line that the preceding team has written and adds a second line. The signal is given and the papers rotate again—each time the group reads and adds another line.

Teams are encouraged to write what comes to mind, even if it's only their name. They must write something in the time allotted. After eight or ten lines, the papers are returned to their original team. Groups can add a line if they choose, revise, and edit the poem they started. The poems can, then, be read orally with team members alternating reading the lines. Later, some of them can be turned into an optic poem (creating a picture with computer graphics using the words of the poem) or acted out using ribbons or penlights (while someone else reads the poem).

WRITING POETRY IN SMALL COLLABORATIVE GROUPS

The collaborative writing of poetry intertwines process with content and students with learning. The cooperative linking of poetry concepts will often turn mundane work into poems rich in detail, sentiment, and humor. The importance of an audience for poetry will help at every stage of the writing.

Working in cooperative groups helps students become more responsible in communicating their understanding to other group members and an audience. The sharing of ideas helps each child develop a better understanding of the writing process and stimulates student conversations around literary pursuits.

When poetry is fused with collaborative dreams, emotions, and comedy, it can foster personal and intellectual growth. If poetry is viewed as a solitary and dour undertaking, then, little space is left for the role of humor in explaining life's goofy splendors.

A group's interactions encourage building and changing ideas to foster the development of collaborative poetry. Interactions between peers can amplify the process. It's good to stir things up, but, sometimes, a few rules will help the group to be more productive:

1. One person should not do all the talking.
2. Accept everyone's ideas.
3. Stick to the topic.
4. Remind each other of the rules or appoint a group leader to help.

Poetry is more than just printed words on a page. Poetry comes alive when the reader and the words connect in a way that provides meaning and build upon the reader's experiences. External stimuli, like some of the methods presented here, can build on sights, sounds, thoughts, and tensions to create poems. Unsaid, inner meanings can be revealed in the "music" or rhythm of a poem. Often, poetry happens between sensibility, control of language, and rhythm.

A good place to begin is with a subject that offers a sense of metaphysical possibility. How do real writers write? Beethoven, for example, would write fragments in notebooks that he kept beside him; later, he would develop these themes. He got ideas from every conceivable direction—including other composers, folk music, and myths. Like other writers, he needed a thorough knowledge of the language (music) and a broad range of experience to build upon. Children can do the same thing by keeping a notebook of ideas about experiences, books, and how their thinking changes in regard to different subjects.

Beyond having some mastery of language, being able to think in images is certainly useful (Bogard and Donovan, 2013). So is the ability to concentrate. Poetry, like any kind of writing, requires many revisions along the way. It's important to get something—almost anything—down and go from there. Sketching out an idea and developing it into a clear vision can foster language growth and help illuminate the reading and writing process (Ganske, 2013).

Since what you read influences how you write, there is a natural connection between a student's written poetry and the richness of the literature program they are exposed to. By leading students to appreciate literature and poetry across time and cultures, teachers can enhance a child's ability to write.

A variety of children's literature and poetry can become a source of vocabulary, metaphor, and conceptual material. Experiencing the language and rhythm of good poetry gives students the building blocks for creating their own poetic patterns. By exposing students to various kinds of poetry patterns, teachers enable them to do a better job of creating poetry on their own.

EXPERIENCES WITH POETRY

1. Poetry with Movement and Music

Poems can be put to music and movement. In a darkened room, one student can read the poem and the rest of the group can move around and using their penlights. You might want to have them practice their movements first (with the room lights on) before presenting to the class. The darker you can get the space for the final presentation, the better. If you don't have a dark enough room, you can use silk scarves or streamers.

Students can also illustrate picture books to go with poems that they can later share with younger children.

2. Daily Oral Reading of Poetry

Students sign up and read poetry aloud at the end of each day. Other students "point," commenting on parts of the poem that catch their attention. A classroom anthology of poetry can be illustrated and laminated.

3. Responding during Free-Writing Period

Thirty minutes to an hour and a half is set aside each day for students to write on any topic, in whatever form they choose. A share time follows so that other students may respond to each other's writing by pointing and asking questions.

4. Literature Share Time

Students gather in small groups once a week to share books they have been reading. The groups are structured so that each student

a. reads the author and title of each book;
b. tells about the book;
c. reads one or two pages aloud; and
d. receives responses from members of the group, specifically pointing out parts they liked and asking questions.

5. Wish Poems

Each student writes a wish on a strip of paper. The wishes are read together as a whole for the group. Students, then, write individual wish poems that are shared.

6. Group Metaphor Comparisons

Poems containing metaphors are read aloud. Group comparison poems are written on the board. Students write individual comparison poems and share them with the class.

7. Sample Poetry Lesson

A lesson developed from *Dinosaurs,* a poetry anthology for children edited by Lee Bennett Hopkins.

a. Teacher reads poems aloud.
b. Students brainstorm reasons why the dinosaurs died, and use words that relate to how the dinosaurs moved.
c. Models of dinosaurs and pictures are displayed and talked about.
d. Students write poems and share them.

8. Cinquain Poetry: Five Lines Long

- Find a picture (that shows action) that you would like to write about.
- Discuss what you would like to write about it with a partner.
- Think of a story that you would like to tell.
- Count syllables as you write your words or phrases to tell a story.

    1. Two syllables in the first line.   *Title*
    2. Four syllables in the second line.   *Description*
    3. Six syllables in the third line.   *Action*
    4. Eight syllables in the fourth line.   *Feeling*
    5. Just two syllables in the last line.   *Conclusion*

- *Write, revise, and share.*

9. Shape Poems

Words should be written to show the shape of the thing being expressed. Try to make a picture out of the poem and add some color to outline it.

slippery   *slithering*   *snake.*
      sliding   *sensitive*

10. The Fame to the Name

This activity can be adapted to suit class and curriculum needs, in integrating whole-language, cooperative learning, and poetry. It is a simple lesson with lots of flexibility that integrates social studies, history, and mathematics. The form of the poem is in the name. The name chosen is written vertically on a chart pad or on the board.

Beginning with each letter in the name, the class brainstorms a sentence or phrase that tells something about the name. Here is an example using the state of Alaska:

A lot of fresh air
*L*and of the froze
*A*thabaskans, Eskimos, Aleuts
*S*eals, bears, moose, and more
*K*aleidoscope in the night sky
A state to be proud of.

Model the alternate line reading of a poem with the class: you read one line, one of your students reads the second line, and so on, to the end of a poem. This will give the children an idea of how it is to be done. Pair students up for poetry, create one or two poems, and practice one poem that each pair presents to the whole class.

Let students know about using alliteration to add excitement. Another idea is to make a list of nouns or adjectives that pertain to the theme. Incorporate cooperative learning by breaking the students into small groups. One method is to have each group work on one letter of the thematic poem or invent their own topic. Be sure to allow time for group presentations.

CREATING POEMS FROM WORDS IN THE ENVIRONMENT

This activity is designed to increase students' observation of words in their environment and create poetry from printed words they observe around them. This can be in the classroom, at school, on field trips, at the bus stop, or walking down the street.

Pair children up or have them grouped in threes. Set the physical boundaries, limiting them to the classroom, hallway, playground, and so on. Set expectations based on the needs of the class.

Language and Literacy

SUMMARIZE WITH BIOPOEMS

Biopoems encourage students to make inferences and synthesize by selecting the precise language to fit the form and character. Biopoems are an effective way to write and think about people in the past and present. They are also effective in introducing students to each other.

Creating a Collaborative Biopoem

1. Divide the class into groups of two or three.
2. Each partnership produces a biopoem on an author of a book, a historical figure, or someone in the news. (Students may have to look up some of the information.)
3. Give each pair of students a large sheet of paper and some colorful markers. Instruct the groups to make a poster large enough for the whole class to see.
4. After the group has finished one biopoem, encourage individuals to do a biopoem about themselves.
5. At each stage, group members should be talking to each other and comparing their work. Select one of the poems and be sure it is reproduced in big enough letters so that the rest of the class can see it.

The biopoem form is this:

Line 1 First name _____
Line 2 Four traits that describe the character _____
Line 3 Relative (brother, sister, friend, parent, etc.) of _____
Line 4 Lover of _____ (three things or people)
Line 5 Who feels _____ (three items)
Line 6 Who needs _____
Line 7 Who fears _____
Line 8 Who gives _____
Line 9 Who would like to see _____
Line 10 Resident of _____
Line 11 Last name _____

Biopoem uses:

- to review a character in a book.
- to understand a historical or well-known figure.

- to introduce people.
- as an activator to develop a character before writing a story.

NEWSPAPERS AND WORLD KNOWLEDGE ARE IMPORTANT

Newspapers can be an excellent supplement to literature, original documents, and oral histories. With the upper grades, middle school, and high school students, newspaper articles can spark ideas for group discussion and provide writing models for analysis.

Students can see how a composition is organized as they read, rather than watch television. They can also compare the evening news, which is often based on items in national newspapers with written stories. The *New York Times*, the *Washington Post*, the *Los Angeles Times,* or the local newspaper can be more stimulating than textbooks. With younger children, simple pieces (with pictures) and publications like the *My Weekly Reader* (Scholastic) can replace more difficult newspapers.

The daily newspaper, particularly if it's in a second language, can be an intimidating document for students to tackle. It is imposing in format and vocabulary for early readers who are accustomed to materials geared toward their competency levels. By preparing imaginative exercises using a newspaper or a clip from a national newscast, a teacher can provide an introduction and demystify those pages filled with newsprint and connect to a second-language video segment.

It is important that the newspaper and video segment cover some of the same ground. The TV news items or conversations should be shown first so that what the students have listened to (and seen) is, then, applied to print. This means using print and video material from the same stories.

The following example called *The Newspaper Scavenger Hunt* is an exercise that can be applied to a variety of reading levels. A list is drawn up with columns of words and phrases extracted from a sample paper. This list of cartoons, pictures, words, concepts, and short phrases (to be found in the newspaper) is handed out to pairs of students along with the paper. They are, then, asked to begin the hunt: Students put the number of the page on which the item is located on the answer sheet and then circle the item in the newspaper. A time limit is set for the searches. When time is up, the students can compare their "success" rates. This exercise can be modified for a range of ability levels.

The teacher can go over the newspaper with them as the class collaboratively searches for connections with the nightly news program the students are asked to watch as homework. Small groups can also make up a creative story composed almost entirely of headlines, subheadings, and a few connections of their own. Political cartoons, with their words removed, can also be presented and groups can come up with their own caption.

INTEGRATING THE LANGUAGE ARTS WITH READER'S THEATER

Reader's theater is the oral presentation of prose or poetry by two or more readers. Complete scripts can be provided or students can write them after reading a story or a poem. The actual story or chapter may be ten or twenty pages long—the finished reader's theater script may only be two or three pages. We recommend trying some prepared scripts first (so that children get the basic idea) and then have the students work as a small group to transform a story or poem into a script.

The typical reader's theater lesson involves script writing, rehearsal, performance, and follow-up commentary for revision. Before the class presentation, children need a chance to practice and refine their interpretation. Everybody eventually gets their own copy so that they can read their role from a handheld script. (A few mistakes in reading are good for a hearty laugh.)

When reading, they stand up (from a chair) or turn to face the audience; when their turn is over, they sit down or turn their back to the audience. If there are four roles and five children, then, two read the same thing at the same time; if there are four students and six roles, then, two members of the group assume two roles.

Lines, gesture, intonation, and movement are worked out in advance. Individual interpretations are negotiated between group members. The performance in front of an audience can intensify the experience and connect the reader to the audience. After going over a prepared script or two, children can take a story that they are reading and create their own scripts.

Reader's theater can be a good informal cooperative learning activity where students not only respond to each other as character to character but in spontaneous responses that ties the group together with the situation of the text. The idea is to use a highly motivating technique to engage children in a whole range of language activities and literate behaviors.

A sample script:

No chairs or turning around for this one. Students jump *up* when their reading turn comes and *down* when they are not reading. If it says "EVERYONE," then, the whole group jumps up to read and down at the same time until it's over. "Put some real energy into it or you get to do it twice." (Time for practice first could help.)

Song of the Popcorn

EVERYONE: Pop, Pop, Pop!
1st child: Says the popcorn in the pan!
EVERYONE: Pop, Pop, Pop!
2nd child: You can catch me if you can!
EVERYONE: Pop, Pop, Pop!
3rd child: Says each kernel hard and yellow!
EVERYONE: Pop, Pop, Pop!
4th child: I'm a dancing little fellow!
EVERYONE: Pop, Pop, Pop!
5th child: How I scamper through the heat!
EVERYONE: Pop, Pop, Pop!
6th child: You will find me good to eat!
EVERYONE: Pop, Pop, Pop!
7th child: I can whirl, and skip, and hop!
EVERYONE: Pop, Pop, Pop, Pop! Pop, Pop, Pop!

EVERYONE should be sure to quickly go up and down with each "Pop. Pop, Pop!" If there aren't enough children for a script, someone can take on two roles. If there are too many students, two can take the same role at the same time.

You can connect the reader's theater popcorn activity directly to the STEM subjects by actually popping popcorn and measuring the distance it goes when the top is off. (You might want to put something on the floor first.) Students can go online first to find the science behind the popping. Hint: it has something to do with a tiny drop of moisture inside the kernel turning into steam.

COMBINING POETRY AND READER'S THEATER

Poetry is an art form that allows us to think deeply about ourselves and others. It doesn't have to be solitary, in the schools, and off the streets. Wordsworth may have been wrong when he explained poetry as "emotion recollected in tranquility." Teachers can always be looking for ways to make poetry interactive and stretch the possibilities. Reader's theater is one way to make poetry inclusive and fun.

Students can divide up a poem in several ways, read each version out loud in their small group, and discuss which version lent dramatic effect to the piece. Other questions: Which script has the best logical breaks or shifts? What are the dramatic or thematic advantages to the different arrangements? How would different interpretations of what the poem is saying affect the division? If there are four students and three roles, two students can read in unison.

Script #1

Reader 1: "Abandoned Farmhouse" by Ted Kooser

He was a big man,
says the size of his shoes on a pile of broken dishes by the house;
a tall man too, says the length of the bed in an upstairs room;
and a good, God-fearing man, says the Bible with a broken back—
on the floor below the window, dusty with sun;
but not a man for farming,
say the fields cluttered with boulders and the leaky barn.

Reader 2:

A woman lived with him,
says the bedroom wall papered with lilacs and the kitchen shelves
covered with oilcloth,

Reader 3:

and they had a child, says the sandbox made from a tractor tire.

Reader 2:

Money was scarce,
say the jars of plum preserves and canned tomatoes
sealed in the cellar hole,
and the winters cold, say the rags in the window frames.
It was lonely here, says the narrow gravel road.

Reader 1:

Something went wrong, says the empty house
in the weed-choked yard.
Stones in the fields say he was not a farmer;

Reader 2:

the still-sealed jars in the cellar say she left in a nervous haste.

Reader 3:

And the child?
Its toys strewn in the yard like branches after a storm—
a rubber cow, a rusty tractor with a broken plow,
a doll in overalls. Something went wrong, they say.

"After-Reading" Questions

- Why do you think the characters left?
- Does this reader's theater version of the poem help you get to know more about the characters?
- Does each reader contribute equally?
- What are the dramatic or thematic advantages to this division?

Script #2

Reader 1: "Abandoned Farmhouse" by Ted Kooser
Reader 2: He was a big man,
Reader 4: (echoes) a big man
Reader 1: says the size of his shoes on a pile of broken dishes by the house;
Reader 2: a tall man too,
Reader 4: (echoes) a tall man
Reader 1: says the length of the bed in an upstairs room;
Reader 2: and a good, God-fearing man,
Reader 3: (echoes) good and God-fearing
Reader 1: says the Bible with a broken back
on the floor below the window, dusty with sun;

Reader 2: but not a man for farming,
Reader 1: say the fields cluttered with boulders and the leaky barn.
Reader 3: A woman lived with him,
Reader 1: says the bedroom wall papered with lilacs
and the kitchen shelves covered with oilcloth,

Reader 4: and they had a child,
Reader 1: says the sandbox made from a tractor tire.
Reader 3: Money was scarce,
Reader 1: say the jars of plum preserves
and canned tomatoes sealed in the cellar hole,

Reader 2: and the winters cold,
Readers 3 and 4: "oh, so cold,"
Reader 1: say the rags in the window frames,
Reader 3: It was lonely here,
Readers 3 and 4: "so lonely"
Reader 1: says the narrow gravel road.
Everyone: Something went wrong,
Reader 1: says the empty house in the weed-choked yard.
Reader 2: Stones in the fields say he was not a farmer;

Reader 3: the still-sealed jars in the cellar say she left in a nervous haste.
Reader 4: And the child? Its toys are strewn in the yard
like branches after a storm—a rubber cow,
a rusty tractor with a broken plow, a doll in overalls.

Everyone: Something went wrong, they say.
Reader 4: (echoes) Something went wrong.

Questions:

- Are there any dramatic or thematic advantages to this division? Explain the thinking behind your preferences.
- How would your conception of what happened to the characters influence how you would divide the poem up?

CREATIVE DRAMA IN THE LANGUAGE ARTS CLASSROOM

Creative drama can serve as a tool for integrating language learning experiences. It also offers teachers a medium that can make important contributions to children's literacy development (Jensen and Nickelsen, 2013). Engaging in literacy-related creative drama should be part of every language arts program. These activities can be done in small groups with few props, no memorization of lines, and no chance for failure.

Dramatic play can be used to bridge the gap between written and visual forms of communication. For example, students can work in small groups to script, act, and even videotape a one-minute commercial. They pick a topic, develop a skit, practice, and perform it for the class. They can, then, critically examine the reasoning behind each group's presentation to the class. The original commercials developed by the students can also be compared to those done on the Internet or on television.

Creative drama can help students reconstruct their own meanings as they respond to literature, writing, and ideas. The way students are asked to go about this process influences their development as readers, writers, and thinkers.

Creative drama

- doesn't emphasize performance.
- adapts to many types of books, lessons, and subjects.
- encourages the clarification of ideas and values.

- evokes contributions and responses from students who rarely participate in "standard" discussions.
- can be used to assess how well students understand what they are reading—characterization, setting, plot, conflicts, and so on.
- provides a stimulating pre-writing exercise.

TEACHING STORY DRAMATIZATION AND TEAMWORK

1. Select a good story—and then, tell it to the group.
2. With the class, break the plot down into sequences or scenes that can be acted out.
3. Have groups select a scene they wish to dramatize.
4. Instruct the groups to break the scene or scenes into further sequence, and discuss the setting, motivation, characterizations, roles, props, and so on. Encourage students to get involved in the developmental images of the characters—what they did, how they did it, why they did it. Have groups take notes on their discussions.
5. Meet with groups to review and discuss their perceptions. Let them go into conference and plan in more detail for their dramatization.
6. Have the whole class meet back together and watch the productions of each group. Instruct students to write down five things they liked and five things that could be improved in the next performance.
7. Let the players return to their groups at the end of all group performances and evaluate the dramas using the criteria applied in number 6.
8. Allow groups to bring back their group evaluations to the whole class. Discuss findings, suggestions, and positive group efforts.

Creative Drama with Active-Learning Teams

1. Personification

This can also be used as a pre-writing activity.

Each student draws the name of an inanimate object (pencil sharpener, doorknob, wastebasket, alarm clock, etc.). Students pick a partner and develop an improvisation.

2. Using Drama to Extend a Story

Creative drama can extend a story. Try "blocking" a play as you read it aloud in class. Giving such a visual perspective increases concentration.

3. Increasing Research and Journalism Skills

Using techniques of role-playing and creative drama, have student groups show *how* to interview. (Give good and bad examples.) Short excerpts from TV news or radio information programs provide good models for discussion and creative drama activities.

Creative drama has long been used to help students learn speaking, listening, thinking, and social skills. After they reflect the theme and action of drama, children can compare character development in plays and written literature. By drawing on prior knowledge and tapping so many modes of expression, creative dramatics ties in nicely with Howard Gardner's ideas about multiple pathways to learning.

Creative dramatics can be a motivating foundation linking all of the language arts by engaging students in the full range of language expression. Reading, writing, and the other language arts often occur in less of a social context. As students experience puppetry, pantomime, improvisation, and other dramatic activities, they develop teamwork skills that can be applied to a wide range of comprehension strategies.

Are innovative leaps in language and the arts different from those in science and engineering? Try making a Venn diagram of overlapping circles with these subjects. Similarities are where they overlap and differences are where the subjects have their own part of the circle. It's a little difficult with four circles, so using two subjects at a time is an alternative.

ALGORITHMS AND NATURAL LANGUAGE GENERATORS

(See if you can tell which paragraph under this subtitle was written by a robo-writer.)

It has been said that language is what makes us human. Well, if that were ever true, it's less true now. Algorithms and natural language generators are getting faster and a lot better.

Revision is the key to good writing—and right now, robo-writers don't seem as good at that as humans.

Automated Insights is the robo-writer program that the Associated Press uses, cutting a lot of newspaper jobs in the process. Two examples of robo-writer programs that we like are *Narrative Science* and *Wordsmith*. Tell these software programs your intended audience and what writing style you prefer. You can even choose the style of a particular author if you wish. Then, your robo-writer will mine a tremendous amount of data, sort out what's useful, and quickly provide you with an original report—or an entire book.

Big data may provide too much information for humans to sort through, analyze, and quickly write about. But would turning the sorting and *composition process* over to computer algorithms cause us to miss out on some of the things that the curious and inventive human mind might come up with?

E-BOOKS AND E-READERS

Critics like Nicholas Carr have looked at the Internet and argue that the Web has had a detrimental effect on language, literacy, and thinking (Carr, 2014). Might digital technology dictate human literacy acquisition and thought processes in detrimental ways?

Amazon's Kindle Readers have helped make e-books ubiquitous. There are differences between reading on your computer (slower) and reading on one of the Kindle devices (faster).

In spite of a little early criticism, the technology now allows for the constant upgrade of nonfiction works. Also, it is sometimes possible to customize the information you want and upgrade articles. Special features let you adjust the size of the print, the sound, or the background lighting. And unlike paper books, e-books can be read in the dark.

Electronic books are able to search for strings of ideas or words within the text and quickly find information, interesting passages, and references. A menu lets readers jump quickly to a particular chapter or a favorite scene.

Some e-book software can even read out loud to you in ways that reinforce the speaking and listening dimensions of language and literacy. As you might imagine, children are very attracted to these digital reading tools. A common question when they see one is this: "When can I read my books on one of these?"

In spite of many possibilities, many students and teachers prefer the elegance of the printed work on paper. Some believe that print will be diminished in importance in a sea change from paper to electronic books. Don't count on it. Paper books will *not* go the way of the quill pen.

At school and at home, both old and new electronic media are increasingly reshaping the way language and literacy acquisition takes place. Although much of language arts instruction will continue along traditional paths, schools cannot avoid the most powerful information and communication technologies of our time. Caution at the primary level is one thing, but complete avoidance in the upper grades is quite another.

The more powerful our digital gadgets become, the more important it is to connect the technology to language and literacy learning—as well as higher-level thinking. Artificial intelligence will not replace original human thought.

But digital technologies will work together with people to extend their reach. So, it is more important than ever to design applications and devices that work well with users.

OLD AND NEW MEDIA FOR LANGUAGE LEARNING

When it comes to using digital technology at the elementary school and middle school levels, the evidence of usefulness is thin. It's a different story with high school students. But for everybody, at least one thing is for sure—on the Internet, multimedia is pushing text-based information to the side of the road. Teachers should not assume, however, that the new thing is always better than the old thing. E-books and online reading devices may have found a comfortable niche, but there is something elegant about print and illustrations on paper.

In spite of inevitable changes, the wood-pulp business has little to fear. Paper documents have proved more resilient than expected. In fact, print on paper is doing much more than hanging on. So, many new avenues for paper documents continue to develop—so much so that we are using more paper than ever.

Books in the traditional mode will be with us throughout the twenty-first century. They are, after all, a user-friendly medium that will not become unreadable when the technology changes. Best of all, you can share a real book or article with a friend. (Recent laws like the Digital Millennium Copyright Act make online sharing of digital materials more difficult legally.)

People who use specially designated software to read an e-book are identified every time they read something. If you aren't authorized to read a specific e-book, you are committing a crime. Fortunately, we have the freedom to read and share paper books. And it is still possible to go to the library.

Since e-books and reading online are part of our future, we had better consider them in social and educational contexts. There are some advantages to carrying fifty pounds of books on an e-reader. Although they are not superior to printed books, the e-book opens a range of possibilities for helping readers master words, passages, and concepts in a new way. Stumble over a word that you can't pronounce or a concept that you don't understand. The e-book will give you the correct pronunciation and an explanation.

Electronic media will not become the only container of content. E-books and digital assistants (artificial intelligence) are simply pieces in the language and literary puzzle. As far as the Internet is concerned, text is fading from the scene as sounds and images become the universal language. Visual literacy is becoming more important than ever.

Digital tools can help us comprehend both the printed word and imagery. Still, for any foreseeable future, paper newspapers and printed books will continue to help with learning to read, write, watch, and investigate.

It helps if students learn how to use digital devices to leverage "big data" and cloud computing. New media offers the possibility of providing new experiences and new approaches to language and literacy learning. Virtual reality also has its place: as many other digital tools, it can enhance existing learning environments.

An array of digital products will continue to evolve as they spread across the landscape. An exotic collection of literary hybrids will change how we read, write, and communicate. Books and magazines can be reduced to digital pieces where readers only have to pay for the pages or chapters that they want to use. Already, many books have become part of distant databases, allowing readers to extract and combine what they want from pools of digital information.

CHANGING THE WAY WE COMMUNICATE AND THINK

A few large tech companies are accelerating the harvesting and centralization of data in ways that redefine how we relate to just about everything in the world. Algorithms are changing what we see and the way we read. Also, it is now possible to chop different books up into interchangeable parts and provide a highly targeted text.

Facebook, Twitter, and Google News are three examples of search engines and social media that customize what you read with mathematical formulas (algorithms). Whether it is book passages, newspapers, or magazines, predictions about what you might want to read (based on prior choices) are constantly being placed in front of you.

It's a vague new world of information fragments arranged by code and delivered online. Although the Internet can expand the range of things to read, social networks like Twitter and Facebook tend to connect you with people who share your point of view. Automation may help or hinder our work across the curriculum. The trick is to avoid the dull, the surveillance, and the manipulation.

Using personalized algorithms and artificial intelligence, it is possible to sort out some of the things that you want to see. At the moment, AI doesn't do as well with creative problems requiring a lot of judgment as a knowledgeable human. So, in the classroom, it is never a good idea to allocate control over what is going on to the latest software or digital gadgets.

Both old and new media are important things. But they are far from being the only things. Many new ideas have been found at the intersection of art and science. But as far as the classroom is concerned, nothing beats a teacher who understands the characteristics of effective instruction.

Digital texts have transformed what and how many of us read, write, and communicate. Still, it is a mistake to assume that the new media is automatically better than the old. In fact, polls have shown that even digital natives strongly prefer the printed page for pleasure and learning. (See Baron's *Words Onscreen: The Fate of Reading in a Digital World*, 2015 book.)

Whether it is a book printed on paper or an e-book, the key to quality language and literacy instruction is a language-rich classroom with a caring and capable teacher. If you want to use the latest technology, be sure to put the pedagogical and curriculum piece in place first. The next step is figuring out how the digital tools can help you achieve instructional goals.

In and out of school, completely avoiding the most powerful information and communication technologies of our time is not an option. But buyer beware. There are plenty of entrepreneurs out there who would like to sell all kinds of tech things. If you let them, they would be happy to impose their version of the future on you. A better arrangement is for educators, policy makers, and the public to exert some control over the vested interests and the rough edges that surround new technologies.

At least one thing is certain—the more powerful technology becomes, the more indispensable good teachers are.

SUMMARY, CONCLUSION, AND LOOKING AHEAD

Language and literacy learning are, at their best, a shared and expanding set of experiences. With parents with younger children, this means stories at bedtime, a discussion of the news, adults who read, museum excursions, and library visits. At school, it means encouraging a cooperative environment so that students can actively construct knowledge together.

Active team learning provides students with opportunities to jointly interpret and negotiate meaning—making connections between prior knowledge and new ideas. Starting with the child's own experiences and background knowledge, the collaborative process can lead to both a shared group idea and a more elegant individual expression (Blauman, 2018).

The interpersonal and integrated nature of literacy development gives students new tools for using and reflecting on a wide range of communication possibilities. By sharing knowledge among peers, the seeds of literacy can grow—allowing students to adjust the language medium that they are using so that they can communicate effectively with a variety of audiences (Bainbridge et al., 2012).

As children develop the different language skills needed to communicate about problems, they also gain the intellectual tools they need to create with others. Engaging youngsters in an active group exploration of ideas is an exciting and powerful way for children to take an active role in their learning community. The ultimate goal is to understand and influence the direction of change in their world.

A major goal of language and literacy instruction is to enable young people to make sense of the world around them. The specific methods that a teacher chooses to teach reading, writing, and other language skills will probably reflect a unique combination of professional knowledge, policy requirements, personal choice, and passion for the subject being taught.

As you navigate the shoals of doing too little and daring too much, it is important to remember that good teachers take surprisingly different paths. Still, *informed enthusiasm* is a trait that all successful teachers have in common. To paraphrase William Blake, *energy is an external delight.*

Uncertainty rules in today's ambiguous and quickly changing world. But whatever happens, you can be sure that change favors enthusiasm, energy, teamwork, and *the prepared mind.*

REFERENCES

Ashton, K. (2015). *How to Fly a Horse: The Secret History of Creation, Invention, and Discovery.* New York: Doubleday.

Bainbridge, J., Heydon, R., and Hibbert, K. M. (2012). *Constructing Meaning: Teaching the Language Arts K–8.* Scarborough, ON: Nelson College Indigenous.

Baron, N. S. (2015). *Words Onscreen: The Fate of Reading in a Digital World.* New York: Oxford University Press.

Blauman, L. (2018). *The Inside Guide to the Reading-Writing Classroom.* Portsmouth, NH: Heinemann.

Bogard, J., and Donovan, L. (2013). *Strategies to Integrate the Arts in Language Arts.* Huntington Beach, CA: Shell Education.

Borges, J. (1957). *Book of Imaginary Beings.* New York: Penguin Classics.

Carr, N. (2014). *The Glass Cage: Automation and Us.* New York: W. W. Norton.

Catmull, E. (2014). *Creativity, Inc.: Overcoming the Unseen Forces That Stand in the Way of Inspiration.* New York: Random House.

Costigan, A. (2018). *An Authentic English Language Arts Curriculum: Finding Your Way in a Standards-Driven Context.* New York: Routledge.

Dierking, C., and Jones, S. (2014). *Oral Mentor Texts: A Powerful Tool for Teaching Reading, Writing, Speaking, and Listening.* Portsmouth, NH: Heinemann.

Dole, J., Donaldson, E., and Donaldson, R. (2014). *Reading Across Multiple Texts in the Common Core Classroom, K–5. [Common Core Standards in Literacy.]* New York: Teachers College Press.

Ganske, K. (2013). *Word Journeys, Second Edition: Assessment-Guided Phonics, Spelling, and Vocabulary Instruction.* New York: Guilford Press.

Jensen, E., and Nickelsen, L. (2013). *Bringing the Common Core to Life in K–8 Classrooms: 30 Strategies to Build Literacy Skills.* Bloomington, IN: Solution Tree Press.

Johnson, D. (2014). *Reading, Writing, and Literacy 2.0: Teaching with Online Texts, Tools, and Resources, K–8.* New York: Teachers College Press.

Hopkins, L. B. (1987). *Dinosaurs.* Boston, MA: Houghton Mifflin Harcourt.

Kooser, T. (1980). *Sure Signs: New and Selected Poems* (Pitt Poetry Series). Pittsburgh, PA: University of Pittsburgh Press.

Moline, S. (2012). *I See What You Mean: Visual Literacy K–8.* Portland, ME: Stenhouse Publishers.

National Council of Teachers of English (NCTE) and International Reading Association (IRA). (1996). *Standards for the English Language Arts.* Urbana, IL: NCTE; Newark, DE: IRA.

Pace, G. (1991). "When Teachers Use Literature for Literacy Instruction: Ways That Constrain, Ways That Free." *Language Arts* 68 (1) (January): 12–25.

Rief, L. (2018). *The Quickwrite Handbook: 100 Mentor Texts to Jumpstart Your Students' Thinking and Writing.* Portsmouth, NH: Heinemann.

Serravallo, J. (2018). *Reading and Writing Narrative.* Portsmouth, NH: Heinemann.

Stock, P., Stock, T., and Schillinger, T. (2014). *Entering the Conversation: Practicing Literacy in the Disciplines.* Urbana, IL: NCTE.

Taylor, B., and Duke, N. (Editors). (2014). *Handbook of Effective Literacy Instruction: Research-Based Practice K–8.* New York: Guilford Press.

Willingham, D. (2017). *The Reading Mind: A Cognitive Approach to Understanding How the Mind Reads.* San Francisco: Jossey-Bass.

Vygotsky, L. S. (1978). *Mind in Society: The Development of Higher Educational Processes.* Cambridge, MA: Harvard University Press.

Chapter 6

Arts Education

Connections, Knowledge, and Informed Encounters

> Reductionism is the distillation of larger scientific or aesthetic concepts into smaller more tractable components. These components can be used by artists and scientists alike to pursue their respective truths.
>
> —Eric Kandel

Whatever the medium, everyone can profit from learning how to analyze, interpret, and create meaning in artistic work. The arts not only enrich our lives; they enhance our understanding of the world around us. Artists have always used whatever tools were available to them. But it is important to realize that aesthetic ideas (or life) do not always come by way of a screen. Concrete physical activities and personal collaboration matter.

Beyond providing new ideas, joy, and entertainment, the arts can challenge thinking and structures of power. Whether it is the visual arts, music, dance, theater, or the media arts, there are many ways that aesthetic experiences can provide important insights and help make us full human beings.

The arts can enrich the lives of eager and reluctant learners. For example, students at any level can relate artistic ideas (from music to the visual arts) to personal experience in ways that help them understand the societal, cultural, environmental, and historical context within which they find themselves.

Complex computer-based mathematical systems are now reaching a point where they can open doors to all kinds of knowledge and creative possibilities. We now have digital devices that can edit this page, create visual art, and compose original music. Artificially intelligent software can create new kinds of art, while serving as tools for new kinds of artists. For now, humans may be better when it comes to issues of the "heart" like dreaming, empathy, and reflection—to say nothing of the ability to capture diffuse feelings of alienation.

In spite of their limitations, digital devices, the Internet, and artificial intelligence (AI) are now part of the fabric of everyday life. AI techniques, for example, are growing better and better at creating poetry, visual art, and sounds that no one has ever heard. Also, more and more of our personal and public expressions have been captured and stored online.

Whether they are high-tech or low-tech, the arts can serve as a connection within and across disciplines. It can also serve as a prism that allows students to connect with multiple subjects, dimensions, and directions of focus (Diaz and McKenna, 2017). Investigation, synthesis, interconnection, and generating a sense of community often come with the territory.

The arts, like other subjects, is often part of a composition that is independent of the whole in which it participates. There is no freedom in an intellectual vacuum. What's going on in the world around you is bound to influence your work. Whether it is the visual arts, music, dance, or drama, there is always the possibility of generating entry points to inventive thinking, innovation, and global skills. The challenge is building on that potential in ways that make some of the possibilities a reality.

The aesthetic quality of everything we create and interact with has an effect on our thinking and well-being. We never think alone (Sloman and Fernbach, 2017). To paraphrase Dieter Rams, individuals build on a community of knowledge to produce well-executed designs—which are the only road to imaginative ideas and beautiful objects.

> There are inescapable links between science, art, and community.
>
> —da Vinci

ENGAGING EAGER AND RELUCTANT LEARNERS

At school, engaging methods of teaching the arts can spark students' imagination and open up inventive possibilities (Douglas and Jaquith, 2018). Critical thinking, creativity, and collaboration are all related parts of a process cutting across a wide range of subjects and real-world challenges. A sample of questions:

- How was the art or music made or composed?
- Can you project yourself into the painting or musical composition?
- What decisions did the artist or composer have to make in their work?
- What might you have done differently if you were in the position of the artist?

The arts provide unique problem-solving possibilities that often trigger unpredictable changes in other areas. Innovation is often a by-product of imaginative efforts to solve a particular problem. The value of education in the arts can also be justified on the basis of its distinctive value in human life as a tool in preparing for unpredictable change.

New media can sometimes be a powerful amplifier for arts education. A good example is how images of visual art from museums around the world can be studied and analyzed online. Digital technology can also influence music education by making sampling and blending music relatively easy by providing students with the digital tools needed to mix and edit the compositions of others as they add pieces of their own compositions (McPherson and Welch, 2018).

In an arts-rich classroom, learners come to appreciate what happens at the intersection of art and technology—a powerful tool to engage learners in subjects across the curriculum. You can paint, compose a song, develop a dramatic script, or dance about just about anything. To paraphrase Isadora Duncan, *If I could write about it I wouldn't have to dance it.*

Along the path to understanding and creating, instruction in the arts energizes students as they develop the skills to create, adapt, and take risks in the future (Chatterjee, 2013). To paint, sing, dance, or take part in creative dramatics are all good examples of activities that enable young people to learn how to deal with the world of today and fashion the world of tomorrow.

THE ARTS AS AN OPENING TO THE FUTURE

Imagination is to break through the limits.

—Henry David Thoreau

The merger of globalization and information technology requires that we all adapt quicker, work smarter, and better understand the natural world. Along with helping students deal with a connected world, the arts can enhance awareness of the aesthetic qualities of their own environmental surroundings.

National content standards in arts education are playing an important role in quality control and supporting efforts to develop assessment techniques to measure student achievement in the arts. All curriculum frameworks for teaching the arts assume that *all* students—not just the gifted—deserve high-quality experiences in this area. Like science or math, you do not have to be aiming at becoming a scientist or mathematician to profit from developing some understanding of the subject matter.

The arts have a lot to do with enshrining some reproduction of experience, gaining some control over the process, and influencing the future. Encounters with the arts also have a unique capacity to provide openings for imaginative breaks from the expected. They continue the universal human practice of making *special* certain objects, sounds, movement, or representations that have been linked with human survival for countless generations.

The arts are showing more potential than ever for enriching technology, science, the environment, and the world around us. In the classroom, they can challenge students to integrate what they are learning—while widening and deepening their imagination. Clearly, if the arts are missing from daily life, this deficiency opens doors to the danger that is inherent in dividing science from humanism.

Real artists pick up hints of cultural and technological challenges well before they result in transforming changes. Using new media to create art is but one example that students could study and emulate. Universally acclaimed for his use of paint on canvas, Hockney has always liked experimenting with old and *new media*. He has made use of paint, photography, iPads, computers, fax machines, photocopiers, and opera set design. In every case, the medium has always influenced—and often determined—the end result.

A good digital example is his use of the iPad and the *Brushes* app. For his iPad drawings, he uses a stylus; you can see the results and in the arrangement of pixels from a color inkjet printer. He likes to turn digital play with his iPhone and iPad into onscreen images and paintings (Hockney and Gayford, 2016; APTN, 2018, Sep. 17). He has even called the iPad the "most spontaneous medium" he has ever found. He has used digital gadgets to capture images like landscapes, plants, sunrises, and sunsets (Hockney and Gayford, 2016; APTN, 2018, Sep. 17). This highly regarded artist has exhibited his digital art work in museums. He also likes spontaneously drawing informal "iPictures" and sending them to friends.

You can use some of Hockney's approaches in the classroom. With the help of two elementary school teachers, we tried one of his techniques with fifth and sixth grade students. He calls the method we used a "joiner."

Our approach: students were put into teams of two or three and asked to take pictures of the same thing from different distances and slightly different angles. For example: students used cameras to take seven or eight pictures of each other and/or things in the environment. The next day, the teacher brought in the prints, students made an arrangement, and glued or taped them on construction paper. [Some of the students liked laminating the end result.] Finally, the results were posted in the room and in the hallway.

Some teachers preferred to have student teams use digital cameras to compose a "joiner" and printed the pictures (immediately) right in the room. The same art construction procedure was followed, but the lesson went from start to finish in one class period. [Note: some students started by going online to see how Hockney did it.]

PROVIDING A SENSE OF OPENING

The arts have always provided a space, a sense of opening, a loving of the question, and a unique communal resource. In addition, the work of artists has often refashioned what is around them in a way that provided creative sparks for the cultural conversation. Today's school reform process should not push aside such a basic aspect of social consciousness and interdisciplinary knowing. If there are no arts in a school, there are fewer alternatives to exploring subjects by the spoken and written word.

Extending education in the arts with other subjects must go hand in hand with other new aspects of schooling and daily life. The notion that the arts can encourage wonder, inquiry, speculation, and technological literacy has for too long been lost in a morass of indifference, nostalgia, crafts, didacticism, and an already overcrowded curriculum. To dig it out requires a greater emphasis on professional development to help teachers become more familiar with the arts and discipline-based art education.

There is an increasing trend for teachers around the world to be asked to encourage students across a number of art forms. The art forms teachers deal with are tools that include: painting, photography, music, drama, media, technology, dance, and performance (Sinclair et al., 2008). These and other art forms give students the opportunity to engage meaningfully with cultures from around the world of today and the world of yesterday (history).

Whatever waves of change sweep over the schools, performance, creation, and understanding will continue to be important. However, arts education is becoming a little more focused on analysis, history, and culture. In the field, this is referred to as a "discipline-based approach." It depends more than ever on the intellectual preparation and commitment of the teacher. Specialists can help, but it is the regular classroom teacher who will continue to be the primary source for instruction in the arts.

FROM DREAMWORKS ANIMATION: A TABLET FOR YOUNG CHILDREN

DreamWorks and a technology company (Fuhu) have a product we have used with five-year-olds; it is called *DreamTab*. The two companies worked together to create a device that has original content and frequent updates.

For example: DreamWorks animators have designed interactive drawing lessons for young children. Some of the best animators in the country use their approaches in their art form. Along the way, young children are invited to experiment with related art of their own. A major goal is to change the way children interact with technology.

You can arrange it so when children turn on the DreamWorks tablet, up pops the penguins from *Madagascar* to get children to do a creative dance. It is about as far as you can get (on a tablet) from the days of solitary confinement in front of a computer screen. Still, it is important to remember that there are frequent times when it is best to turn off the devices and work in small groups—or *go outside to play.*

Although most of the experiences are at a primary level, the DreamTab comes with a stylus that is similar to what professional animation artists use. It is also possible to get on the Web and do things like e-mail and instant messaging. Fortunately, everything is designed to fit within the guidelines set by the Children's Online Privacy Act.

A NATIONAL SECURITY AGENCY (NSA) WEBSITE FOR CHILDREN

Like DreamWorks, the NSA uses animation to put a smiley face on its mission. No, it is not to inform on their parents—at least not yet. Children are encouraged to construct their own secrets using furry animals and reptiles. Breaking colorful codes and recovering signals from around the world are part of the game.

Decipher Dog and *CryptoCat* are just two of the cartoon characters who help children learn about spying duties and think about what they want to do when they grow up. After entering the *How I Can Work for the NSA* section of the site, a bunny rabbit comes along and says that it likes listening to hip-hop and rock music. In his free time, he uses cutting edge technology derived intelligence from a variety of signals from all over the world.

The NSA animation takes children into games of identifying the purpose, content, and user of the site. Cell phones, e-mails, Google searches, Facebook records, and other signals are all part of the fun. The CyberTwins (Cy and Cyndi) even have some good tips on stopping and thinking before sharing private information on social network sites. The whole process resembles efforts by many businesses, educators, and other governmental agencies to promote self-serving messages and related interaction with children. *A note to parents and teachers:* you had better keep an eye on what youngsters are doing on the Internet.

THE ARTS—PAST, PRESENT, AND FUTURE

Even at the beginning of human civilization, the arts had a central place in ceremonies that connected cave paintings to ritual, religion, and daily life. In fact, humanity has been shaped by the synthesis of science, math, art, and the imagination (Gurche, 2013).

In the future, will our culture be as filled with the arts as it now is with television and sports? Of course, a lot depends on how you narrow or stretch the definition of "the arts." But no matter the definition used, the arts and social change will continue to shape each other.

Both the arts and the sciences have many things in common; both are always looking for interesting problems to solve. Making or exploring something that we are unfamiliar with is the essence of creativity across subjects. And we all have a certain amount of creativity in each of us.

Many educators and artists suggest that children without knowledge of the arts are as ignorant as children without knowledge of literature or math (Eisner, 2005). Schools are good at transmitting knowledge, but a deep knowledge in the arts requires going beyond factual information to developing practical knowledge. In a paint-by-numbers world, you have to know when to depart from the cookbook. This involves avoiding tunnel vision and building on imitation, imagination, and actual experience to create things that go beyond the textbook or lesson.

De Tocqueville predicted that American democracy would diminish the character of art. It took decades to prove him wrong—at least for a while. Now, some artists have worked hard to break down the disconnection between the nation's establishment (including the arts, academia, and the press) and the people.

Whether it's visual arts, music, or the theater, the arts have the potential to help us be receptive to new thinking and generous toward the production of something fresh. Far from being beaten down, American artistic expressions, especially in film and music, have been some of our most successful exports.

Some people think of the arts as elitist, therapeutic, frivolous, impractical, or even mindless entertainment. They are not always wrong, but they miss the point. The arts can provide important intellectual tools for understanding many subjects. They also build on qualities that are essential to revitalizing schooling: teamwork, analytical thinking, motivation, and self-discipline (Blandy and Bolin, 2018).

In the twenty-first century, the arts have a lot to offer new approaches to curriculum and instruction. For example, skills and perspectives of art may be used to show what thinking, learning, and life can be. The arts also provide cultural resources that people can draw on for the rest of their lives. But in this era of accountability, attention has to be given to the substance of the disciplines involved. Otherwise, the arts may be dismissed as expendable in an era of curriculum gridlock and financial difficulties.

VISUAL ANALYSIS AND THE CRITICAL FUNCTION OF THE ARTS

Efforts are now being made to deepen and extend education in the arts by connecting them to critical thinking, problem solving, aesthetic analysis, technology, and new ways of working. This increasing influence of discipline-based art education (DBAE) curriculum addresses more than the traditional issues of creative expression and performance. It provides an interdisciplinary framework for connecting arts education to aesthetic criticism within a cultural, historical, and social context.

In a literature-based reading curriculum, for example, students are expected to develop the thinking skills necessary for "literary criticism." Should we expect less when it comes to the arts? A renewed emphasis on artists, criticism, aesthetic discourse, and the importance of discipline-based art education will accompany education into the next century.

DO MUSIC LESSONS HAVE A COGNITIVE BENEFIT?

Cognitive skills like spatial reasoning may be enhanced, but the research on the subject isn't conclusive. But even if future studies rule out major gains in academic achievement, music education generates joy and cultural knowledge along with the development of musical skills.

SOME SELECTED EXAMPLES OF DISCIPLINE-BASED ART EDUCATION ACTIVITIES (PRACTICAL IDEAS FOR TEACHERS)

The following discipline-based art activities are organized around an interdisciplinary unit theme titled "You and Your World." This approach was selected so that critical thinking skills and interdisciplinary content could be linked and included as an integral part of classroom life.

As part of an integrated approach, it is important that children learn to be more flexible, and move freely between different communications media. To accomplish this, children need exposure to many different communication forms.

Unit: Introduction Activity

Before beginning this unit, discuss with children the need all people have to communicate ideas and how there are many ways to do this. Encourage children to brainstorm all the ways people use to communicate. List the suggestions on the board or a chart. Young children may wish to find or draw pictures that can be placed on a bulletin board. Such a chart becomes an ongoing resource for students to refer to, and additions can be incorporated throughout the year.

Unit: You and the World

When you think of how you are related to others, the thing that most people say is family. But even if you were alone in the world, you wouldn't be unrelated. The fact that you have read these words makes you a member of English-speaking people. As a student, you have a relationship with those who attend your school and with those who work there. The music you listen to and enjoy is enjoyed by others. Your relationships with your fellow humans are marked by the foods you think are good, the clothes you think are fashionable, the jokes you tell, and so on.

In the nineteenth century, Ralph Waldo Emerson viewed the relationship between the arts and your day-to-day work with others as central to imaginative thinking. He felt that people depended on their relationships in order to understand what they read, write, paint, or sing. Emerson thought that each person is related to a few others and to all people; each of us has within ourselves the sum of human history.

You may never have thought of yourself as part of an ongoing historical record. But chances are you have watched characters in movies or on television and sensed that they felt as you have felt and acted as you would have acted, that they were, in a sense, related to you. Your relationships with actual or fictional others are the basis of sympathy and one of the keys to imaginative thinking.

Unit Activities

1. Make a map of significant relationships in your life. Put your name in the center of a sheet of paper. Then, begin thinking of the important people in your life. As you think of them, write their names on the paper. Organize or group the names that belong together. You may wish to connect the names with lines to show the relationships.

2. Make a list of ten words you chose at random from the dictionary. Next, write or make up something about you that uses all the words you have listed. It could be a paragraph in the form of a news report, a story, a creative drama—whatever works with the words you have. Just make sure you *use* the words, not just mention them.

For example the word *hare*.
Use: I saw a *hare,* chewing on a carrot in my garden.
Mention: *Hare* is another word for a rabbit.
Let the words guide what you write.

3. Suppose there is a lottery in your state. A three-digit number is picked at random. For $1, you can buy a ticket picking any number from 000 to 999. If the number on your ticket matches the number on the ticket drawn, you win $500. Is that a good payoff? Why or why not? How much of the money the state takes in does it keep?

4. Try reflecting on and then, describing an episode from a television series that you regularly watch. Here are some questions that may help you think about the program. Jot down your answers. Then, write a paragraph or two about what you've learned.

   a. For what sorts of people is the program produced?
   b. Are the main characters people like yourself? Are they people you want to be like?
   c. Are the main characters unusual in some way? If so, in what way? Are they usually attractive? Do they have special skills?
   d. If the program is a comedy, what are the jokes about? Is there a laugh track? Do you laugh as often as you hear people in the audience laughing?
   e. What kinds of problems do the characters in the program have? Are they the same sorts of problems you have?
   f. Are the characters in the program richer or poorer than you are?
   g. Describe the plot. Does it make sense? Do the characters in the program act the way real people act?
   h. Does the program use background music? What sort of music? What does the music contribute to the mood of the program?
   i. Try looking at the program without listening to the sound. What do you notice? Try listening to the program without watching the picture. What do you notice?
   j. Do you know what is going to happen before it happens or are you surprised? How do you feel when the program ends?

CONNECTING SUBJECT MATTER WITH THE ARTS

The arts can also help get a dialogue going between disciplines that often ignore each other. When knowledge from diverse subject matter areas is brought together, the result can be a new and valuable way of looking at the world. The arts and humanities have proved very useful tools for integrating curricular areas and helping students transcend narrow subject matter concerns.

Teachers at many levels have used intellectual tools from the fine arts as a thematic lens for examining diverse subjects. Some schools have even worked out an integrated school day, where interdisciplinary themes based on the fine arts add interest, meaning, and function to collaboration. Whether it's reading, writing, arithmetic, or anything else, the arts can be wrapped around central themes in the arts so that rich connections stimulate the mind and the senses.

The research suggests that using a thematic approach improves students' knowledge of subject matter and aids in the transfer of the skills learned to other domains outside the school. An additional finding is that good units organized around themes can improve the students' ability to apply their knowledge to new subjects.

In collaborative art projects, language development flourishes when children are encouraged to discuss the materials they are using and reflect on the nature of their artwork through writing. Whatever the combination, an important result of integrating various subjects around a theme results in an enhancement of thinking and learning skills—*the meta-curriculum.*

Before we can deal with teaching the thinking process, children need some solid content to think about. After that, teachers need to provide continuity between activities and subjects. The thinking skills engendered in one area can serve as a connection between subjects. In making curriculum connections, it's often helpful for teachers to see model lessons that include cross-disciplinary suggestions and activities.

The relationships established between subjects and the way teachers facilitate these relationships are important. When disciplines are integrated around a central concept, students can practice the skills that they have learned from many subjects. This helps students make sense out of the world.

The goal of an interdisciplinary curriculum is to bring together different perspectives so that diverse intellectual tools can be applied to a common theme, issue, or problem. Thematic approaches can help by providing a group experience that fosters thinking and learning skills that will serve students in the larger world.

Organizing parts of the curriculum around themes means that each subject is mutually reinforcing and connected to lifelong learning. Subjects from the Greek classics to radiation theory need the historical, philosophical, and aesthetic perspective afforded by interdisciplinary connections.

Curriculum integration provides active linkages between areas of knowledge, and consciously applies language and methods from more than one discipline to examine a central theme, issue, topic, or experience. This holistic approach focuses on themes and problems and deals with them more in depth rather than just memorizing facts and covering the text from cover to cover.

There is always the danger of watering down content in an attempt to cover all areas. We can, however, teach the work of Newton on one hand, while paying attention to the history of the times on the other. The history of ideas, political movements, and changing relationships among people are part of the fabric of our world. We cannot narrowly train people in specialist areas and expect them to be able to deal with the multifaceted nature of twenty-first-century jobs.

THEMATIC STRATEGIES FOR CONNECTING SUBJECTS AND PEOPLE

> Artists have the right—and possibly the obligation—to reinterpret the history of our time.
>
> —Oliver Stone

Like the arts, innovation in science can experience ups and downs and cul-de-sacs. These different ways of knowing—the arts and the sciences—do not need to grow further apart. The unity of all cultural and scientific efforts were the unwritten rule until the eighteenth century. But as art and science have progressed over the last 200 years, both have become more narrow, specialized, and extensive.

The arts can help connect the mind and the senses—uniting the cognitive and affective dimensions of learning. They can also encourage group cooperation. Whether the collaboration is in the distant past, on the Internet, or with a small group of peers sitting nearby, teamwork in art can focus on creating, interpreting, and connecting to others. Socially useful art requires *hard* thinking about the location and the intended audience in order to understand how best to engage local modes of expression and needs.

Themes can also direct the design of classroom activities by connecting classroom activities and providing them with a logical sequence and scope of instruction. One set of steps for developing thematic concepts include:

1) Determine what students know about a topic before beginning instruction. This is done by careful questioning and discussion.
2) Be sensitive to and capitalize on students' knowledge.
3) Use a variety of instructional techniques to help students achieve conceptual understanding.
4) Include all students in discussions and cooperative learning situations.

Thematic instruction values depth over breadth of coverage. The content should be chosen on how well it represents what is currently known in the field and its potential for dynamically making connections.

THEMATIC UNITS

The design of thematic units brings together a full range of disciplines in the school's curriculum: language arts, science, social studies, math, art, physical education, and music. Using a broad range of discipline-based perspectives can result in units that last an hour, a day, a few weeks, or a semester. They are not intended to replace a discipline-based approach, but act as supportive structures that foster the comprehensive study of a topic.

Teachers can plan their interdisciplinary work around issues and themes that emerge from their ongoing curriculum. Deliberate steps can be taken to create a meaningful and carefully orchestrated program that is more stimulating and motivating for students and teachers. Of course, shorter flexible units of study are easier to do than setting up a semester or year-long thematic unit.

Collaborative thematic curriculum models require a change in how teachers go about their work. It takes planning and energy to create effective integrated lessons and more time is often needed for subject matter research because teachers frequently find themselves exploring and teaching new material. Thematic teaching also means planning lessons that use nontraditional approaches, arranging for field trips, guest speakers, and special events.

Contacting parents, staff members, and community resources who can help expand the learning environment is another factor in teachers' time and planning efforts. Long range planning and professional development for teachers are other important elements of the process.

The arts have a power beyond aesthetics for making us "see." They can also enhance the ability (flexibility) to change your mind in the light of new information. The arts can help us view ourselves, the environment, and the future differently—even challenging our certainties about the arts themselves. In connecting the basic concerns of history, civilization, thought, and culture, the arts provide spatial, kinesthetic, and aesthetic skills that are the foundation of what it means to be an educated person. Such understandings do not occur spontaneously. They have to be taught.

The process of understanding or creating in the arts is more than unguided play, self-expression, or a tonic for contentment. They can be tools for shattering stereotypes, changing behavior, building a sense of community, and as a vehicle for socio-political commentary. An example from the visual arts:

Barbara Kruger develops popular imagery that merges words and concepts from other disciplines. Along with other post-modernist artists (like Keith Haring and Jenny Holtzman), she works outside the artistic and the aesthetic frame to harness the formative power of images to affect deep structures of personal and social belief.

In a similar manner, artist Alexis Smith combines quotes, flotsam, and jetsam that speak to the artifices and pitfalls of a mythical America. When the right object is connected to the perfect quote, the result can range from the humorous to the toughest and most intriguing social observation.

Moving toward music, storytelling, and dance, Lori Anderson extends the edges with performance art, combining nearly every basic art form with literary references and video imagery to create theatrical performances. Like many modern artists, she releases possibilities by making use of collaborators across time, media, and subject matter.

MUSIC LESSONS AND SUCCESS ACROSS THE CURRICULUM

Participation in music lessons may lead to higher levels of achievement in other subjects. An edited version of a STEM music lesson developed by Victoria Abbott:

The Five Senses Lesson Plan:

In kindergarten classrooms across the Maritimes, teachers often teach a unit titled "Exploring the World Using Our Five Senses." The lesson can be linked to a grade 1 unit that is also about the senses: "Exploring Objects and Materials with Our Senses."

Grade Level: Kindergarten or Grade 1
Subject: Science + other subjects.
Unit: "Exploring the World Using Our Five Senses" or "Exploring Objects and Materials with Our Senses."
Introduction: To build on multiple ways of knowing (multiple intelligences), Victoria Abbott suggests using two songs that she has rearranged from songs that children may know.
"Intelligences" include: Musical/Rhythm, Verbal/Linguistic, Bodily/Kinesthetic, and Visual/Spatial.

1) The 5 Senses Songs {Note: The square parentheses are suggested actions for you and your students.}

1) The Senses Family (Sung to the tune of the first verse of "The Addams Family Theme Song")

I have 5 different* senses [Hold up five fingers]
That help me every day
They help me in many ways
Now sing them all with me.
With my eyes I see, [Point to eyes]
My nose lets me smell, [Point to nose]
My ears help me hear, [Point to ears]
And with my tongue I taste. [Point to tongue]
Now, don't forget the fingers; [Wiggle all your fingers]
They help us to feel
Everything we touch [Gently rub fingers together]
Now let's begin again.
I have 5 different senses [Hold up five fingers]
I see, I smell, I taste, [Point to eyes, then nose, and then stick out tongue]
I hear and I feel. [Point to ears and then rub fingers together]
Thanks for singing them with me.

* "Different" should be pronounced "diff'rent" to help the song flow smoothly.

2) There Are 5 Senses (Sung to the tune of "Do Your Ears Hang Low")

There are 5 senses [Hold up 5 fingers] Can you tell
   me what they are? [Shrug shoulders]
Do you use your eyes to see near or far? [Point to your eyes then motion near or far]
Does your nose tell you if something smells good or bad? [Point
   to your nose and then give thumbs up or down]
How many senses do you have?
There are 5 senses [Hold up 5 fingers] Can you tell
   me what they are? [Shrug shoulders]
Do you use your tongue to taste a chocolate bar? [Stick out tongue]
Do your fingers tell you if something is soft or hard?
Tell me how many senses there are.
Which sense is left? [Shrug shoulders] Can you tell me what it is?
We know you see and smell, taste and feel, [Point to
   your eyes, nose, tongue, and fingers]
We won't forget the last sense, have no fear. [Wag
   your fingers during "have no fear"]

We use our ears so we can hear. [Point to ears]
There are 5 senses [Hold up 5 fingers] Yes, we learned them all.
But did you know there are some people who don't have them all.
Some can't see, feel, smell, or taste. Some can't hear me.
But we are all special and we are unique.

Activity:

[The same procedure can be followed with both songs]

1) Create a big poster with the lyrics written or typed on it.
2) Create 5 smaller posters with each of the senses on them. You can either draw the body parts that are attached to the senses or find pictures to use.
3) Read the lyrics to children while doing the actions.
4) Sing the song to children. (You can play an instrumental in the background, sing it without music, or play along with an instrument.)
5) After the song is done, ask the students to share with a partner what they thought about the song. Depending on the class's reaction, you can decide how often to do the song with the children.

Assessment: An informal assessment can be done by keeping a note on who was able to point to the correct body parts along with the songs.

—Victoria Abbott

OPENING UP A SHARED SENSE OF WONDER

There are connections among productive citizenship, academics, and the arts. For students to make these connections, it will take more than a specialist in the art class for one hour a week or an inspirational theater troupe visiting the school once a year.

Brief experiences can help and inspire—but it takes more sustained work in the arts to make a real difference. Quick "drive-by teaching" is the equivalent of driving a motorcycle through an art gallery; you might get some blurred notion of color but not much else. Avoiding arts education denies students a vital quality-of-life experience—expression, discovery, and an understanding of the chances for human achievement.

The arts can open up a sense of wonder and provide students with intellectual tools for engaging in a shared search. This won't occur if children are having fewer experiences with the arts at school and in their daily lives. They at least have to know enough to recognize what to notice and what to ignore.

This means that some grasp of the discipline is required if the arts are going to awaken anyone to the possibilities of thoughtfulness, collaboration, and life.

There are some excellent models or prototypes of art education. The Minneapolis discipline-based art program is one example. Another is in Augusta, Georgia, where the National Endowment for the Arts (NEA) has supported the development of an exemplary arts education model. This program uses the arts to improve academic achievement, the general learning environment, student self-esteem, attendance, creative thinking, and social equity among students.

ART ACTIVITIES WHICH ENCOURAGE REFLECTION

Reflecting is a special kind of thinking. Reflective thinking is both active and controlled. When ideas pass aimlessly through a person's mind, or someone tells a story that triggers a memory—that is not reflecting.

Reflecting means focusing attention. It means weighing, considering, choosing. Suppose you want to drive home; you get the key out of your pocket, put it in the car door, and open the door. Getting into your car does not require reflection. But suppose you reached in your pocket and couldn't find the key. To get into your car requires reflection. You have to think about what you are going to do. You have to consider possibilities and imagine alternatives.

A carefully balanced combination of direct instruction, self-monitoring, and reflective thinking helps meet diverse student needs. The activities suggested here are designed to encourage higher order thinking and learning and provide a collaborative vehicle for arts education.

1. Looking At the Familiar, Differently

Students are asked to empty their purses and pockets on a white sheet of paper and create a face using as few of the items as possible. For example, one case might be simply a pair of sunglasses, another a single earring representing a mouth, a third could be a profile created by a necklace forming a forehead, nose, and chin. It gives students a different way of looking at things. It's also an example of a teaching concept known as aesthetic education.

2. Collage Photo Art

Students at all levels can become producers as well as consumers of art. We used a videotape of David Hockney's work from *Art in America*. Hockney, one of today's important artists, spoke (on the videotape) about his work and explained his technique. Students, then, used cameras to explore Hockney's photo-collage technique in their own environment.

Student groups can arrange several sets of their photos differently—telling unique stories with different compositions of the same pictures. They can even add brief captions or poems to make more connections to the language arts, social studies, science, or music. Photographers know the meaning of their pictures depends, to a large extent, on the words that go with them.

Note: teachers do need to preview any videos before they are used in the classroom because some parts may not be appropriate for elementary school children. Teachers can also select particular elements and transfer them from one device to another so that only the useful segments are present on the tape used in class.

3. Painting with Watercolors and Straws

In this activity, students simply apply a little suction to a straw that is dipped in tempera paint. Students, then, gently blow the paint out on a sheet of blank paper to create interesting abstract designs.

4. Creating Paintings with Oil-based Paints Floating on Water

Working in groups of three, have students put different colored oil-based paints on a flat dish of water. Apply paper. Watch it soak up the paint and water. Pull it out and let it dry.

[We sometimes use acrylic paint.]

5. Lesson title: Color Mixing

[Version of a lesson plan by Jasmine Roberts.]

The goal is to help students understand which two colors need to be mixed to make purple and green.

Question for the teacher before the lesson: *When and why might it be important to know about mixing colors?*

Student Directions:

Show which two colors mix together (to make purple and green) by painting a picture.

Procedures:

1. Have students work in pairs.
2. Allow each pair to have four colors to choose from to make purple and green.

3. Advise each student to paint something that they can take home and put up.
4. Ask each pair to answer the question: Who might find it important to mix colors?

Objective:

Introducing students to color mixing and establishing the relevance of mixing colors together.

Grade level:

K–1

Materials:

paint, canvas or paper, paper towel, paintbrushes, and aprons.

EXPANDING SOCIAL AND PERSONAL VISIONS OF THE ARTS

The arts are organized expressions of ideas, feelings, and experiences in images, in music, in language, in gesture, and in movement (Eisner, 2005). Teachers can create a space for the arts to flourish—a sense of opening—that helps free students from the predicted and the expected. Using the arts to inquire and sense openings results in what Emily Dickinson called *a slow fire lit by the imagination*. As America moves toward the new millennium, we need all the imagination we can get.

Advancing the understanding of culture, art, creativity, and human values has everything to do with the life and quality of this nation. Nevertheless, educational decision makers often don't pay much attention to these issues. In the United States, for example, the arts are most often found on the fringes of the curriculum and instruction. This is due, in part, to not having a long tradition of broadly prizing artistic expression beyond the cute and the comfortable. Little is expected of citizens or leaders when it comes to knowledge about artistic forms.

The arts can open new horizons, enrich the spirit, and help educate students to expand cultural visions. An artistic perspective can color the way we see other aspects of social and educational change. When the arts are viewed as a personal luxury—and not traditionally associated with "real wage earning" occupations—developing or maintaining a good arts education program is more difficult. This is a disappointing portrait of ourselves, a reflection not of human strength and aesthetic vision, but of their absence. Restoring faith in the arts—and arts education—means expanding the margins to restore faith in ourselves as a nation.

Human societies have always depended on the arts to give insight into truths, however painful or unpopular they may be. Today, in many countries, there is wide agreement that the arts can aid children in developing creativity, becoming good citizens, and being productive workers. The basic notion is that the person and the world are poorer without the arts.

A country's richness of knowledge, enlightenment, and enduring resources for thoughtfulness also benefit from artistic endeavors. From Asia to Europe, serious arts education is one of the integrating features of the school curriculum. Such an investment in the arts is seen as an investment in the community—and vice versa. Americans are beginning to take notice.

Inventing the future of arts education means expanding the links within the arts, the community, and the schools. There is a world out there that students must explore with the arts if they are to be broadly educated—to say nothing of developing self-examination, critical thinking, and problem-solving skills. All of these qualities can be taught and reinforced through the arts. They can also help children integrate thinking skills by such activities as producing critiques, reflecting on aesthetic concerns, and dealing with the nature of our humanity.

Children and young adults frequently have the innate ability to do creative work in the arts. What's frequently missing are basic artistic understandings and the opportunity for expression and analysis. Experience within a discipline matters because it is hard to do something new unless some of it is automatic.

When students have the chance to express themselves, there is the excitement of producing in their own way—conveying their personal aesthetic experience through the use of figurative language (metaphors, similes, etc.) in their writing and symbolism in their painting. The challenge is to provide the necessary background and opening doors so that meaningful concepts and images will emerge.

INCLUDING ART EDUCATION IN SCHOOL REFORM

In an effort to make arts education part of the national curriculum reform, a series of *Discipline-Based Art Education* reports has been put forward by the Getty Foundation. These reports encouraged the schools to help students go beyond crafts to art criticism, history, and aesthetics.

In some of the small-scale projects, art educators, historians, philosophy professors, and local teachers gathered to collaborate in making aesthetics less mysterious for children and young adults. It was felt that even at early levels, students need to be grounded in the ability to reflect on art, study the discipline, and test out the skills involved in production.

The United States provides an example of a national effort to make sure that the arts touch every classroom. The National Endowment for the Arts (NEA) has an agreement with the U.S. Office of Education to create an "in-depth arts-in-education program" that could be part of the effort to "reinvent" American schools (*www.nationalartsstandards.org,* 2016). (National Core Arts Standards, 2018).

The arts are recognized as representing a body of knowledge—as well as a practical study of technique. Isolated school experiments are proving that there are a number of ways of doing this beautifully on a small scale. The question is whether the call for "world-class standards" in the arts will mean real change for a significant number of schools.

Although the connection to a rich artistic tradition is important, no response should be considered *the* "right" one. In fact, seeking the rewards of what some adults see as good creative products often makes their appearance less likely. Instead, teachers can mix modeling intellectual stimulation with the natural rapport and creative production that is such an important part of the mysterious art of good teaching.

Art criticism, history, and aesthetics contribute to production and a child's ability to draw inferences and interpret the powerful ideas. Art (like film, reading, or mathematics) makes use of certain conventions and symbol systems to express figurative meaning. In the visual arts, for example, this may include symbols in its expression through style (the fine detail), composition (arrangement of elements), and by creating the possibility for multiple meanings. "Reading" an artist's symbols is as much of a skill as reading print or video images.

Art means going beyond the transient messages that are often overvalued by the culture. In multicultural societies like Canada, art also means weaving artistic material (visual arts, movement, and music) from other cultures into the curriculum, enabling students to creatively confirm the truth and beauty of their heritage.

Art is not limited to specific times or cultures. Greek art learned from Egypt. Christian art was shaped by ideas from Greece and the East. African, Chinese, Egyptian, and Mexican art have influenced modernism. A high-quality national culture can provide a unifying frame for a rich multiplicity of cultural influences (Gelineau, 2004).

Exposing children to a variety of artistic forms and materials will make it easier to locate areas of strength and weakness. All students may have a similar range of choices, but it is how these choices are made that count. Choosing from a variety of artistic and intellectual possibilities is beneficial for building both the strength of creativity and basic skills. In addition, the arts can also help to get a dialogue going between groups or disciplines that often ignore each other.

When knowledge from diverse subject matter areas is brought together through art, the result can be a new and valuable way of looking at the world. Children can be involved in artistic interdisciplinary projects—ranging from illustrating their own books to designing movement, to poetry, and to producing videos with camcorders. Process, production, and critical dimensions are all important. To understand literature, for example, children must function as critics. With art experiences, critical analysis is equally important.

The creative effect of questioning, challenging, and aesthetic reflection contributes to creative habits of mind and sets up possibilities for action. It is also important for students to see how the arts can set up possibilities for positive action and take on our world concerns. The Art Institute of Los Angeles, for example, was asked to provide design concepts and tools to help solve problems like lack of affordable housing, attractive parks, small shopping centers, and ways to make the community more aesthetically pleasing.

STRATEGIES FOR ARTS-CENTERED COLLABORATION

For teachers: getting students to actively collaborate in thematic lessons that build on the arts requires a depth of planning, modern assessment techniques, and cooperative classroom management skills. Collaborative learning values differences of abilities, talents, and background knowledge. Within a classroom that values teamwork, many conventionally defined "disabilities" integrate naturally into the heterogeneity of expected and anticipated differences among all students.

Organizing an interdisciplinary lesson around a theme can excite and motivate all students to actively carry out projects and tasks in their group. "Disabilities" and "differences" come to constitute part of the fabric of diversity that is celebrated and cherished within cooperative groups. In such an educational climate, no individual is singled out as being difficult and no one student presents an insurmountable challenge to the teacher when it comes to accommodating a student with special needs.

In a collaborative classroom, no student needs to be stereotyped by others when they realize that there are many and varied "differences" among students. It is easier for the student with special needs to fit in. For some pupils, "differences" may in fact constitute a "disability," defined as the inability to do a certain life or school-related task. Such a difference, however, need not constitute a handicap as cooperative learning is a joint enterprise. Some may have a disability or special talent, but all have information and skills to contribute to the learning of others.

The central question is, How do individual classroom teachers, already overwhelmed with tasks, find ways to: adapt collaborative techniques, plan thematically, and modify approaches for successfully accommodating all students in their classrooms?

Collaborative group learning is a proven way to learn how to imaginatively solve problems. Adapting techniques and modifying current methods can help—especially when it involves rethinking the structure of the curriculum and seeking different approaches for teaching students in a way that builds on their unique human qualities.

Art flourishes where there is a sense of adventure.

—Alfred North Whitehead

CONNECTING TO MODELS OUTSIDE OF SCHOOL

Fostering creativity in the arts means encouraging students to think for themselves, coming up with different solutions to problems by linking arts education to their own personal experiences. Just as it is in life, outside of school creativity involves innovative answers to questions that can sometimes change the very nature of the question itself (Bertram and Forbes, 2014).

Creating an educational renaissance will require all the community resources educators can connect with. Some schools are experimenting with residencies by area artists. Others have connected to adult models by sponsoring projects on sites (an art gallery, symphony hall, the ballet company studio). In-depth thematic units can be developed that allow students to work on-site to solve real-world and complex problems, understand subject matter in depth, and make connections across disciplines.

Getting students interested in a topic or problem and interacting with others in an environment that allows thoughtful and creative expression are objectives that few educators will disagree with. Yet how, with today's already cluttered curriculum, testing requirements, and red tape, does a teacher find time to unearth art topics of interdisciplinary interest? Team training can help to share the load and community resources can free up some teacher time. But to keep reform going, we are going to have to change organizational structures and protect teachers from bureaucratic requirements.

Teachers can supply classroom vignettes about effective teaching: the butterfly that "hatched" from a chrysalis in their classroom, students' creative language experience stories, movement (dance), creative dramatics, and painting murals. Other teachers might recall the newscast of the whale trapped in the ice, which spawned an array of activities: research on whales, letters to elected representatives, a bulletin board charting bird migration patterns, and an attitude survey graph. Good teachers know that to be really excited about a subject, they must really care about it.

The social forces surrounding a field of study and individual talent are important factors in generating (or inhibiting) creativity. As far as arts education is concerned, this means: legitimizing its goals by becoming an active force in educational change, assuming a more aggressive role with "at risk" students, and focusing on the potential of the arts to foster thinking skills and problem-solving abilities.

All social and educational institutions convey messages that can affect creativity and artistic development. Deep questions of value are involved in the kind of models we set and our methods for evaluating artistic products. Art may belong to everyone, but being literate in the subject means being able to understand, critique, and create in a whole array of symbol systems.

It's best to get high-quality instructional experiences and training early on. As children gain more aesthetic understanding, teachers can think of them as participants in the artistic process. As students paint their own paintings, compose music, and collaborate in arranging their own dances, they come to experience the inner nature of how aesthetic creativity develops.

A SAMPLE OF CLASSROOM ACTIVITIES THAT INVITE THOUGHTFULNESS:

Create Writing Partnerships

A common collaborative learning strategy is to divide the partnership into a "thinker" and a "writer." One partner reads a short concept or question out loud and tells what he or she thinks the answer should be. The writer writes it down if he/she agrees. If not, he/she tries to convince the "thinker" that there is a better answer. If agreement cannot be reached, they write two answers and an initial one.

Literature and Movement

Some poems, stories, myths, and ballads are particularly suited to interpretation through movement. Choose one or two students to read while the others respond to the reading with creative movements. Create a magical atmosphere with poetry. Use penlights in a darkened classroom or use colorful ribbons for creative movement that requires group effort and harmony.

While the teacher or one of the children reads, have the other children reflect or enact the poem in movement. Each child can hold a penlight or ribbon to help create an effect.

Improvise Short Original Music Pieces

Students can improvise music pieces and variations on existing pieces, using voices or instruments (e.g., traditional, nontraditional, jazz, rock, or electronic music).

Working With a Partner in the Art Museum

In an art museum, students might focus on a few paintings or pieces of sculpture. Have students make up a question or two about some aspect of the art they wish to explore further—and respond to five or six questions from the list in a notebook or writing pad they take with them.

Possible Questions for Reflection:

- Compare and contrast technology and art as ways for viewing the past, present, or future differently.
- How is the artwork put together?
- How are pictures, pottery, and music used to communicate?
- How did the creator of the visual art image expect the viewer to react or respond? Is the content or subject of the artwork the most important part of it? What else might the artist have wished to produce?
- How does your background affect how you view the message?

Visuals are authored in much the way print communication is authored. How does the author of a picture or piece of sculpture guide the viewer through such things as point of view, size, distortion, or lighting?

- What are the largest or smallest artistic designs of the work?
- What is the main idea, mood, or feeling of the work?
- When you close your eyes and think about the visual, what pictures do you see? What sounds do you hear? Does it remind you of anything—a book, a dream, TV, or something from your life?
- How successful is the sculpture or artwork? What is your response to it?
- Where did the artist place important ideas?
- How do combinations of or different ways of organizing things make you feel?
- Does the artwork tell us about big ideas such as courage, freedom, or war?

- How does it fit in with the history of art?
- What does the work say about present conflicts concerning art standards, multiculturalism, and American culture?
- How did the work make you feel inside?
- Was the artistic work easy or hard to understand?
- Why do you think it was made? What would you like to change about it?

ESTABLISHING A COLLABORATIVE ARTS COMMUNITY

Valuing a range of contributions within a supportive and collaborative community can make the difference between a competent self-image and the devastating belief that nothing can be done "right." Recasting the teacher's role from authority figure dispensing knowledge to that of a collaborative team leader (coaching mixed-ability teams) is a major ingredient of collaborative learning.

Making students active participants in deciding what and how they should learn doesn't diminish the need for informed decision makers. But without these—and other changes—in the power relationships within schools and within the schoolroom, educational reform will be stymied. This process is particularly important with some media (like video) because it often takes a small group to do much of the production.

In a collaborative setting, the teacher helps students gain confidence in their ability and the group's ability to work through problems and consequently rely less on the teacher for validating their thinking. This involves a conceptual reexamination of today's student population, the learning process, decision-making relationships, and classroom organizational structure.

Challenges for the professional teacher in this new environment:

- taking a more active role in serving students of multicultural backgrounds and "at risk" students. In many cases, this means addressing non-Western artistic formats.
- focusing and taking advantage of cooperative learning teams to foster students' thinking, reasoning, and problem-solving abilities.
- making use of cooperative learning strategies, peer tutoring, and new technology to reach a range of learners and learning styles.
- working to professionalize arts education and legitimize the arts in the schools. This includes assessment of student knowledge, ability, and performance.
- developing exemplary materials supportive of cooperative learning.

This development will have to be done with particular attention to: the promotion of thinking skills, the needs of "at risk" students, the needs of teacher professionalism, assessment, accountability, and the advent of new technologies.

Although children are capable of both imitation and figuring out structure on their own, they can use mechanisms for thinking and digging deeply into subject matter and themselves. They also need structures for analyzing works of art, music, dance, and drama, and frameworks for sorting out what is real in the environment. Children have widely divergent talents and interpretations that they derive from their own perceptions and ways they filter the world.

It is difficult to consider products of the imagination apart from the system of values brought to it. Good exercises in art education involve students in altering familiar or unfamiliar images along lines they feel are promising. Students need the chance to try things out, reflect on what they have done, and try again. Most teachers know how to encourage or reorient students if they are getting nowhere.

Good teachers believe all children will learn and recognize the need for high expectations as they strive to reach every individual. Successful instructors are also able to facilitate, probe, and draw on additional information, examples, and alternative approaches for those students who were unable to connect with the information initially. This requires knowing enough about the subject to feel comfortable with it.

All of our students possess the capacity to absorb knowledge—but it takes intelligent teaching to use that knowledge to reason effectively. Curriculum development requires staff development. It is often adult models (like teachers) and family support that make the difference between a commitment to the arts or dismissing them as irrelevant.

Effective teachers strive to ensure what's being learned is a center of interest for students. This often means walking a fine line as they engage students as active thinkers—without interfering when children are working well on their own. Creative experiences in the arts are a blend of informed adult encouragement and opportunities for creative exploration.

A flexible arts curriculum requires not only knowledge about each child but judgment about when to intervene, recognizing (like Emerson) that sometimes it is best to "let the bird sing without deciphering the song."

THE ARTS AS A FRAMEWORK FOR INQUIRY AND INVENTION

There was a time when you could look to artists, musicians, poets, dramatists, and other creators in the arts to arrange things in a way that would enlighten human understanding. Is something missing in the twenty-first century? STEM-related activities can help students integrate the arts in a way that provides students with opportunities to apply the design processes of invention and innovation. Along the way, the arts can help put scientific, social, and political conditions into a creative framework and bring multiple dimensions of problems into view.

Along with integrative art investigations in the classroom, it is important to make sure that all students become confident and competent with the arts. This includes making sure that every child has access to a rigorous arts curriculum in a climate of reasoned thoughtfulness and high expectations. In addition, tomorrow's instruction in the arts will be more discipline based and pay more attention to developing sophisticated consumers of the arts.

The history of pictures, for example, begins in caves and can now be found on computer screens (Hockney and Gayford, 2016). Now, painting, photography, electronic images, and the whole range of art forms are entwined. Who knows where any of the arts will connect or reside in the future?

How might the arts transform reality into imaginative new and encouraging images?

Improving education has as much to do with improving cultural quality as it does with increasing productivity. Much of what students have to do in the world outside of school involves the ability to work in groups, self-regulate, plan, execute, and complete various kinds of work.

When the arts are used as a framework for inquiry across the curriculum, they also serve as a productive way for both teachers and students to rethink and better understand basic real-world issues (Barone and Eisner, 2012). As teachers help their students learn to identify artistic themes, develop questions, and examine works of art, new and exciting pathways can be opened for examining all kinds of topics.

Arts education can be an agent of social change, in general, and education, in particular. If visual artwork, music, dance, and drama are not found in the public schools, then the chances for thoughtfulness, self-expression, and aesthetic appreciation are bound to be diminished. On a broader plane, the arts can help counter the tendency for standardization in the school reform process.

Imaginative behavior involves breaking out of established patterns and looking at things in different ways. Being open to surprise and surprising others are part of the process (Livio, 2017). This involves sharing ideas and knowledge with others. The possibilities the arts offer for a unique opening up of new spaces will be sorely missed if they are relegated to the margins of educational restructuring.

SUMMARY, CONCLUSION, AND LOOKING AHEAD

The visual arts, dance, poetry, plays, and music have long been organizers or points of integration for a whole range of human activities. Opening up the classroom to such powerful motivators involves having students think for themselves and explore ways to reinvent their world. The whole process has a lot to do with actually *doing* art—while paying attention to culture, aesthetics, technology, and how the arts relate to other subjects.

In most areas, imaginative work and innovation are sparked by putting together creative teams in an environment that values experimentation and accepts mistakes as part of the process.

Some have suggested that the arts can help students move beyond popular culture's common assumption that the past is our only future. There are always plenty of new ways to say, compose, paint, and construct. And you do not have to paint like Matisse or compose music like Stravinsky to make worthwhile art. All of us can learn to appreciate and be inspired by the arts.

What about building on the arts to influence the future in a positive way? Do we have any concrete ideas—or do we even know where to look or what questions to ask? In 1910, for example, not Picasso, Einstein, Marconi, or anyone else could envision what a visually intensive global communication device (iPhone) would look like.

Something inventive and new is likely to involve many creators who are involved in constant collaboration. From the light bulb to the devices of the digital age, new products usually don't have a single inventor. Thomas Edison and Steve Jobs may have been the first to bring some of their ideas to the masses; but neither one was the first to experiment with artificial light or design MP3 music players.

Innovation and working out the details require individual genius and cooperative teamwork. The same can be said for getting people to buy something new that they didn't know they needed (Johnson, 2014). Imaginative new approaches require both teachers and students who are good at seeing beyond the immediate task.

A step-by-step engineering mentality has its limitations. The arts and humanities are needed to prepare people with broader horizons and so that they can be better at dealing with an unknowable future. Intellectual tools gathered from subjects as diverse as science and the arts are needed to deal with the problems of today and tomorrow. We must all get ready for situations that are unanticipated, need to be discovered, or simply defined in new ways.

The future is always under construction. Although it's something of an open space with many possibilities, not everything can happen, but many things can. The understandings sparked by analysis, mastery, and values inherent in the arts have a lot to do with getting ready for the world we cannot yet know. As far as the classroom is concerned, informed encounters with the arts is a proven way for helping the young look outside today's reality and cooperatively construct imaginative alternate views of what the future might look like.

The years doors open
like those of language,
to the unknown.
Last night you told me . . .
. . . to think up signs
Sketch a landscape, fabricate a plan
on the double page
of day and paper.
Tomorrow, we shall have to invent,
once again,
the reality of this world.

—Octovio Paz (translated by Elizabeth Bishop)

REFERENCES

APTN. (September 17, 2018). "David Hockney Unveils His iPad Art." *The Telegraph.* https://www.telegraph.co.uk/culture/culturevideo/artvideo/10408677/David-Hockney-unveils-his-iPad-art.html.

Barone, T., and Eisner, E. (2012). *Arts Based Research.* Thousand Oaks, CA: SAGE Publications.

Bishop, E. (1962). *Brazil.* New York: Life World Library/Time Inc.

Bertram, V., and Forbes, S. (2014). *One Nation Under Taught: Solving America's Science, Technology, Engineering, and Math Crisis.* New York: Beaufort Books. [Examines how to help American students who are falling behind in the STEM subjects.]

Blandy D., and Bolin, P. (2018). *Learning Things: Material Culture in Art Education.* New York: Teachers College Press.

Chatterjee, A. (2013). *The Aesthetic Brain: How We Evolved to Desire Beauty and Enjoy Art.* Oxford: Oxford University Press.

Da Vinci, L. (2017). *The Literary Works of Leonardo Da Vinci, Compiled and Edited From the Original Manuscripts: Vol. 1 of 2.* London: Forgotten Books.

Diaz, G., and McKenna, B. (2017). *Preparing Educators for Arts Integration: Placing Creativity at the Center.* New York: Teachers College Press.

Douglas, K., and Jaquith, D. (2018). *Engaging Learners through Artmaking: Choice-Based Art Education in the Classroom.* 2nd edition. New York: Teachers College Press.

Eisner, E. (2005). *Reimagining Schools: The Selected Works of Elliott Eisner.* New York: Routledge.
Gelineau, R. P. (2004). *Integrating the Arts across the Elementary School Curriculum.* Belmont, CA: Wadsworth.
Gleick, J. (2016). *Time Travel: A History.* New York: Pantheon Books.
Gurche, J. (2013). *Shaping Humanity: How Science, Art, and Imagination Help Us Understand Our Origins.* New Haven, CT: Yale University Press.
Harari, Y. N. (2015). *Homo Deus: A Brief History of Tomorrow.* Toronto: Signal.
Hockney, D., and Gayford, M. (2016). *A History of Pictures: From the Cave to the Computer Screen.* London: Thames and Hudson.
Holmes, R. (2014). *The Age of Wonder: How the Romantic Generation Discovered the Beauty and Terror of Science.* New York: Vintage.
Johnson, S. (2014). *How We Got to Now: Six Innovations That Made the Modern World.* New York: Riverhead Books.
Kandel, E. (2016). *Reductionism in Art and Brain Science: Bridging the Two Cultures.* 1st edition. New York: Columbia University Press.
Livio, M. (2017). *Why: What Makes Us Curious.* New York: Simon & Schuster.
McPherson, G., and Welch, G. (Editors). (2018). *Creativities, Technologies, and Media in Music Learning and Teaching.* Oxford: Oxford University Press.
Muggeridge, M. (2003). *The Very Best of Malcolm Muggeridge.* Vancouver, BC: Regent College Publishing.
National Coalition for Core Arts Standards. "National Core Arts Standards." (2018). National Arts Standards. https://www.nationalartsstandards.org/.
Pollman, M. (2017). *The Young Artist as Scientist: What Can Leonardo Teach Us.* New York: Teachers College Press.
Sinclair, C., Jeanneret, N., and O'Toole, J. (Editors). (2008). *Education in the Arts: Teaching and Learning in the Contemporary Curriculum.* Oxford: Oxford University Press.
Sloman, S., and Fernbach, P. (2017). *The Knowledge Illusion: Why We Never Think Alone. New York:* Riverhead Books.
Stone, O. (2004). "Oliver Stone Interview." *Charlie Rose Show*, April 14, 2004. https://charlierose.com/videos/13839.
Thoreau, H. D. (2001). *Henry David Thoreau: Collected Essays and Poems.* Edited by E. H. Witherell. Boston: Library of America.
Thornton, S. (2014). *33 Artists in 3 Acts.* New York: W. W. Norton.
Whitehead, A. N. (2010). *Process and Reality*: *Gifford Lectures Delivered in the University of Edinburgh during the Session 1927–28.* 2nd edition. New York: Free Press.

About the Authors

Dennis Adams is a former elementary school teacher who has taught at the University of Minnesota, University of Maine, and McGill University in Montreal. He did graduate work at Harvard University and has a PhD from the University of Wisconsin. He is the author of more than twenty-five books and more than a hundred journal articles on various educational topics.

Mary Hamm has taught at Ohio State University and the University of Colorado. More recently, she has been teaching at San Francisco State University. She has worked on both math and science standards and has published more than a dozen books and eighty journal articles on these subjects.

www.ingramcontent.com/pod-product-compliance
Lightning Source LLC
Chambersburg PA
CBHW031552300426
44111CB00006BA/288